D1496301

Background Independence in Classical and Quantum Gravity

Background Independence in Classical and Quantum Gravity

JAMES READ

OXFORD
UNIVERSITY PRESS

Great Clarendon Street, Oxford, OX2 6DP,
United Kingdom

Oxford University Press is a department of the University of Oxford.
It furthers the University's objective of excellence in research, scholarship,
and education by publishing worldwide. Oxford is a registered trade mark of
Oxford University Press in the UK and in certain other countries

Published in the United States of America by Oxford University Press
198 Madison Avenue, New York, NY 10016, United States of America

British Library Cataloguing in Publication Data
Data available

Library of Congress Control Number: 2023938931

ISBN 978–0–19–288911–9

DOI: 10.1093/oso/9780192889119.001.0001

Printed and bound by
CPI Group (UK) Ltd, Croydon, CR0 4YY

Contents

Acknowledgments

This book grew out of a thesis written in Oxford in 2016 for the B.Phil. in Philosophy degree. As such, it should be of no surprise that my first and greatest intellectual debt is owed to my supervisor on that program, Oliver Pooley, for his meticulous guidance throughout the first stages of the development of this work. My next greatest debt is owed to Harvey Brown, my D.Phil. supervisor, whose contrarian thought has come to pervade all aspects of my thinking on the philosophy of space and time. ("What would Harvey say about this?" is, surely, a refrain uttered by many young philosophers of physics.)

In addition to my supervisors at Oxford, I owe a great many thanks to a great many other philosophers of physics. In particular, I am indebted to Gordon Belot, Jeremy Butterfield, Adam Caulton, Patrick Dürr, Henrique Gomes, Sean Gryb, Sebastian de Haro, Nick Huggett, Niels Linnemann, Keizo Matsubara, Tushar Menon, Thomas Møller-Nielsen, Ruward Mulder, Brian Pitts, Dean Rickles, Simon Saunders, Nic Teh, Will Wolf, and two anonymous readers for Oxford University Press, for invaluable discussions and feedback regarding background independence.

I'm also grateful to Peter Momtchiloff for his wonderful guidance over the course of the publication of this work, and to Deva Thomas for his assiduous work on the typesetting. Last but most of all, my thanks to Sumana and Aditya, without whom there would (of course) be no living The Good Life.

Acknowledgements

1
Introduction

It has become folklore that Einstein's view that his 1915 general theory of relativity (GR) was set apart from other theories in virtue of its *general covariance* was dashed on the rocks of the Kretschmann objection. This objection, put to Einstein in 1917, maintains that *any* theory admits of a generally covariant formulation, so this cannot be a feature which sets general relativity apart from other theories.[1] Einstein conceded the point—but a view has persisted that there must nevertheless remain a *substantive* sense of general covariance, which plays exactly this role. Since the virtuoso study of Norton (1993), this has become extremely well-known amongst those working in the philosophy and foundations of space and time.

But the debate continues. In the aftermath of Norton's work, focus has shifted from substantive general covariance to *background independence*—but, as Teitel (2019) stresses, this fresh new name belies essentially the same debate. The central change, if any, is that this quality of background independence is now regarded by some authors as being one of the central *lessons* of GR, in the sense that it is a property which (the claim goes) should be manifest in any candidate quantum extension of that theory. (Claims of this form can be found in the works of e.g. Rovelli (2001, 2004) and Smolin (2006, 2008), and, from a somewhat different perspective, in those of de Haro et al. (de Haro et al., 2016; de Haro, 2017a); I will assess them over the course of this work.) But this notwithstanding, there still remain multifarious open questions regarding (i) the definition of background independence; (ii) the relation between background independence and other qualities of GR, such as diffeomorphism invariance; and (iii) the status of both classical and quantum spacetime theories with respect to their background independence.

The purpose of this book is to make some progress in resolving these questions, thereby making headway in advancing our understanding of this elusive notion. The plan is as follows. In Chapter 2, I recall some basics from the philosophy of science which will prove essential in the ensuing. In Chapter 3, I consider a number of definitions of background independence, their relations to other important qualities of spacetime theories (including GR) such as general covariance and

[1] Actually, Kretschmann's original point as put to Einstein is a little more subtle: it is conditional on acceptance of Einstein's point-coincidence argument, proposed by the latter in the wake of the hole argument. For discussion, see (Norton, 1989, p. 818).

Background Independence in Classical and Quantum Gravity. James Read, Oxford University Press.
 DOI: 10.1093/oso/9780192889119.003.0001

diffeomorphism invariance, their applications to theories such as GR, scope, interrelations, and limitations.

The above completed, in Chapter 4, I turn to the background independence of five classical spacetime theories: Newtonian gravitation theory, Newton-Cartan theory, teleparallel gravity, Kaluza-Klein theory, and shape dynamics. The work of the previous two chapters facilitates evaluation of these theories with respect to their background independence (thereby allowing comparisons with GR to be made), appraisal of the definitions themselves, and the bringing to light of novel features of these theories.[2]

In Chapter 5, I first discuss ways in which the above definitions of background independence can be modified in order to apply to quantum theories. This done, I assess the background independence of perturbative string theory, bulk theories in holographic dualities, and loop quantum gravity. Based upon the (quantum extensions of the) definitions presented in Chapter 3, I find, for example, a majority verdict that perturbative string theory and holographic bulk theories *are* background independent—meaning that the absence of this quality cannot be cited as a reason not to pursue the study of these theories (as claimed in e.g. (Smolin, 2008, §5.4)).

In completing this work, to repeat, I hope to clarify the space of possibilities regarding what background independence could amount to (in both the classical and quantum contexts); to assess whether GR is in fact privileged in virtue of its background independence (on any of these ways of cashing out the notion); and to assess whether various prominent theories of quantum gravity are or are not background independent (a matter of present contention). Although I do not come down firmly on the side of one particular approach to background independence (for reasons which I will explain over the course of this work), I hope, thereby, to have stoked the flames for the next many decades of dispute on this topic. In this respect, I view the work as being a footnote to the magisterial contributions of heroes such as Norton (1993), Earman (2006), and Pooley (2017).

Before I begin in earnest, two points. First: in this work, questions regarding whether background independence should be considered a desideratum on any spacetime theory are certainly addressed—but they are not the focus of the project. Rather, I am concerned with the *formal* features of this quality, the extent to which such features are exhibited in notable spacetime theories, and the lessons which appraisals of background independence can teach us regarding those theories. For what it's worth: it seems to me that background independence is better understood as an heuristic on theory construction, rather than as a norm: the notion can be a useful guide in our construction of future theories, but a theory should not be

[2] For example, various symmetries of Newtonian gravity (§4.1.3), different senses of gauge invariance in teleparallel gravity (§4.2.3), and issues regarding the diffeomorphism group associated to Kaluza-Klein theory (§4.3.3).

rejected solely on the grounds of its lacking said quality. It is, in other words, a principle to be used in the context of discovery rather than in the context of justification. Second: a reader might object to my assessments of the background independence of certain theories, on the grounds that such assessments differ from their own. To which I say: please, tell me *your* definitions; I'd be delighted to bring them into the fold.

2
Models and Gauge

Throughout this book, I operate within the broad framework of the *semantic conception* of scientific theories—introduced by authors such as Suppes (1960), and famously endorsed by *inter alios* van Fraassen (1980, 1989)—on which a given theory is associated with a class of *models*.[1,2] In this chapter, I introduce in §2.1 three important varieties of model, before in §2.2 elaborating on how *gauge redundancies* fit into this framework. Discussion at this stage is focussed exclusively upon classical theories; consideration of the quantum is deferred to §5.1.

2.1 Three Classes of Model

For a given theory $\mathscr{T}$, we can take the most general associated class of models to be that of *kinematically possible models* (KPMs) $\mathscr{K}$, which consists of tuples of specified geometric objects.[3] For example, the KPMs of general relativity (GR) are

[1] One should distinguish the claim that a given theory has an associated class of models from the (more controversial) claim that a theory should be *identified* with such a class of models. In this book, I embrace the former, but remain agnostic on the latter.

[2] Van Fraassen identifies a model of a theory as "Any structure which satisfies the axioms of [that] theory" (van Fraassen, 1980, p. 53). In the language of this work, it is natural to identify models in van Fraassen's sense with *dynamically possible models*—or perhaps with the intersection of *dynamically possible models* and *boundary possible models* (see §2.1). In this work, following e.g. Pooley (2013, 2017), I understand the notion of a model in a broader sense, discussed below.

[3] Including tensors, but also (in principle) tensor densities, etc. Though discussion of such non-tensorial objects shall mostly be set aside in this work, for arguments for taking these objects seriously, see (Pitts, 2006, 2010). Relatedly: presumably one would also want ultimately to include non-geometric objects such as spinors into one's specification of the KPMs of one's theories; a presentation of such objects liberated from coordinate conditions is, however, elusive—see (Pitts, 2012).

Indeed, on the distinction between geometric and non-geometric objects: the formal presentation of the concept of a geometric object goes back (at least) to works by Nijenhuis (1952) and Schouten (1954), and was later elaborated on by e.g. Trautman (1962, 1965). For a differentiable manifold M, consider an arbitrary point $p \in M$ and two arbitrary local coordinate systems around p. In order to specify a geometric object on M, one writes down (i) a set of components (a set of N real numbers) in each such coordinate system, and (ii) a well-defined rule relating the components in the one coordinate system to the components in the other (Trautman, 1965, pp. 84–85). (This is not necessarily a tensorial transformation law—for example, those for connection coefficients in general relativity, or for tensor densities, are not tensorial, but are nevertheless still well-defined.) Non-geometric objects lack at least one of the above—Nijenhuis (1952) dubs geometric objects denuded of their transformation rules mere 'objects', and by 'non-geometric objects' I will mean such entities in what follows. For an excellent discussion of what constitutes a 'well-defined transformation rule', see (Duerr, 2021, §3.3)—but, in brief, the essential feature is this: if one transforms (the components of) a given object in a coordinate system A to (its components in) a coordinate system C, that must yield the same result as transforming (the components of) the object in A to some intermediate coordinate system B, and then to C.

Background Independence in Classical and Quantum Gravity. James Read, Oxford University Press.
 DOI: 10.1093/oso/9780192889119.003.0002

picked out by all triples of the form $\langle M, g_{ab}, \Phi \rangle$,[4] where M is a four-dimensional differentiable manifold;[5,6] g_{ab} is a Lorentzian metric field on M; and Φ is a placeholder for the matter fields of the theory.

Classically, a theory $\mathscr{T}$, with KPMs $\langle M, O_1, \dots O_n \rangle$ (where the O_i are geometric objects) comes with a set of *dynamical equations* for the O_i. The KPMs of $\mathscr{T}$ in which the O_i obey those dynamical equations form a subspace $\mathscr{D} \subset \mathscr{K}$, the *dynamically possible models* (DPMs) of $\mathscr{T}$.[7] For example, in the case of GR, only those triples $\langle M, g_{ab}, \Phi \rangle$ the geometric objects of which satisfy the *Einstein equation*[8]

$$G_{ab} = 8\pi T_{ab} \tag{2.1.1}$$

—the central dynamical equation of the theory, which relates g_{ab} to the stress-energy tensor T_{ab} of the Φ[9]—in *addition* to the dynamical equations of the Φ, are DPMs.[10, 11]

[4] Throughout this book, abstract (i.e. coordinate-independent) indices are written in Latin script; indices in a coordinate basis are written in Greek script; semicolons indicate covariant derivatives; commas indicate partial derivatives; and the Einstein summation convention is used. Round brackets around indices denote symmetrization over those indices; square brackets around indices denote antisymmetrization. I also set $G_N = c = 1$.

[5] Note with Earman and Norton that one should avoid making *ab initio* “unnecessary global assumptions” regarding manifold topology (Earman and Norton, 1987, p. 518). That said, when we introduce in Chapter 3 theories which contain ‘fixed fields’, we will be treating theories which are not “local spacetime theories” in the sense of Earman and Norton (1987); the reason for doing so is, of course, that there are conceptual insights to be had in assessing such theories.

[6] One should avoid, at this stage, asserting M to be the *spacetime* manifold, for to do so is to conflate the mathematical model under consideration with the possible world(s) to which that model is ultimately interpreted as corresponding (cf. footnote 11). Indeed, in light of the debate over the hole argument (revitalized by Earman and Norton (1987)), it is *not* necessarily correct to interpret M as representing substantial spacetime at all—though this issue shall be set aside in this work.

[7] There are questions to be asked regarding whether the kinematics/dynamics distinction presupposed in my presentation here is, ultimately, either fundamental or well-founded: this, of course, is a central theme in the writings of Brown (2005). I will address these issues later in this work, but for now let me say simply that drawing the distinction between KPMs and DPMs has become very mainstream in the foundations of spacetime theories, and indeed for me there will be significant payoffs from thinking in this way (as we'll see in the next chapter). Thus, I help myself to this machinery. (For more on the kinematics/dynamics distinction, see (Curiel, 2016; Linnemann and Read, 2021b).)

[8] Barring discussion in footnotes 31, 32, and 35 of §3.3.3, as well as in §5.3, in this work I restrict myself to the case of vanishing cosmological constant Λ. For $\Lambda \neq 0$, the Einstein equation reads $G_{ab} + \Lambda g_{ab} = 8\pi T_{ab}$. Note also that by speaking of the ‘Einstein equation’ rather than ‘Einstein equations’, I am being pedantic: (2.1.1) relates certain geometric objects directly, and hence is only one equation. Replacing abstract with coordinate indices would yield a set of partial differential equations—in which case, ‘equations’ would be proper.

[9] Sometimes, the matter fields Φ and their associated stress-energy tensor T_{ab} are conflated. This is a mistake, and should be avoided—for further discussion, see §3.6.4. Note also that T_{ab} is also a function of the metric field g_{ab} (Lehmkuhl, 2011).

[10] Strictly, independence of these dynamical equations from the Einstein equation depends on the case in question—for discussion of this point, see (Misner et al., 1973, §20.6) and (Brown, 2005, §9.3).

[11] Note that we take (following e.g. (Belot, 2013)) the class $\mathscr{K}$ of KPMs of $\mathscr{T}$ to be the class of all possible *histories* of the fields $O_1, \dots, O_n$ (rather than as e.g. instantaneous states). If a history is an element of the solution set of the specified dynamical equations of the theory, then that model is a DPM. On this understanding, one can proceed to interpret models as corresponding to *possible worlds*—see §2.2.

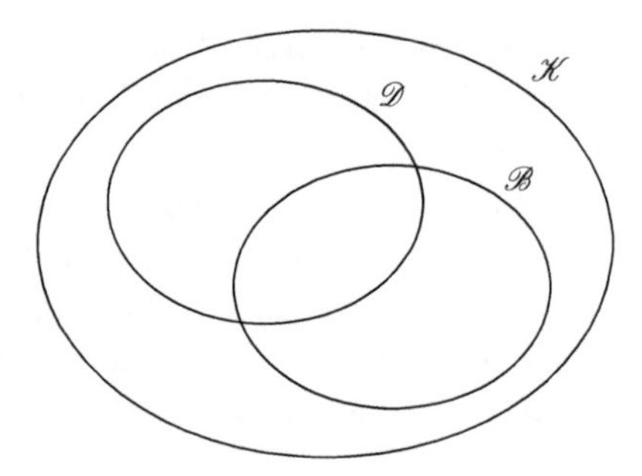

Fig. 2.1 The relation between $\mathscr{K}$, $\mathscr{D}$, and $\mathscr{B}$ for a generic $\mathscr{T}$.

Though the notions of KPMs and DPMs of a theory are known in the literature (see e.g. (Pooley, 2013, 2017)), I introduce here a third class of models of a theory $\mathscr{T}$, which I call the *boundary possible models* (BPMs) $\mathscr{B} \subset \mathscr{K}$. These are those KPMs of $\mathscr{T}$ the O_i of which satisfy certain specified boundary conditions. Again taking the example of GR, sometimes we restrict attention to KPMs of this theory for which the metric field g_{ab} satisfies the boundary condition of *asymptotic flatness* (see e.g. (Wald, 1984, ch. 11), and §§3.5, 3.6.3, and 5.3.1).[12]

Where a class $\mathscr{B}$ is specified for $\mathscr{T}$, we may take those models to specify the *KPMs* of a restricted theory $\mathscr{T}'$. In that case, $\mathscr{T}'$ is a *sub-theory* of the *super-theory* $\mathscr{T}$.[13] Finally, for a given $\mathscr{T}$ with specified dynamical equations and boundary conditions on its geometric objects, there is *in general* (i.e. abstracting from the context of particular cases) no restriction on the overlap between $\mathscr{D}$ and $\mathscr{B}$—the generic picture thus being as per Fig. 2.1.

2.2 Gauge Redundancies

Models of a theory $\mathscr{T}$ may be *interpreted* as representing possible worlds.[14] Sometimes, however, we may wish to interpret two or more distinct models as representing the *same* world. In that case, the space of KPMs $\mathscr{K}$ of $\mathscr{T}$ is partitioned into classes of *gauge-equivalent* models—which are interpreted as representing the same world—and the multiplicity of models representing the same world is a case of a *gauge redundancy*.[15] In the case in which the interpretation of $\mathscr{T}$ leads to gauge

[12] For a recent article exhibiting the use to which the notion of BPMs can be put, see (Wolf and Read, 2023).

[13] The significance of this move will become clear in §§3.5 and 5.3.4.

[14] This move is extremely common—see (Baron et al., 2022) for discussion. That being said, it's questionable whether associating models of theories with complete possible worlds is always appropriate—this is a point made forcefully by Wallace (2022a). For my purposes in this book, these latter worries (with which I am generally sympathetic) won't much matter.

[15] There are other senses of 'gauge' in physics—for example, by a 'gauge theory' is sometimes meant a Yang-Mills theory. For discussion of this in the context of spacetime theories, see e.g. (Wallace, 2015; Weatherall, 2016b).

redundancy, we may construct a *reduced* space of models $\tilde{\mathscr{K}}$, each model in which is (up to isomorphism) associated with a gauge equivalence class of models in $\mathscr{K}$.[16]

The above is purely formal; there remains an outstanding question concerning *when* two models of $\mathscr{T}$ should be interpreted as representing the same world. One popular line (found in e.g. (Saunders, 2003, 2007; Greaves and Wallace, 2014; Dewar, 2015)) is as follows. First, define a *symmetry transformation* to be a transformation upon the O_i in the KPMs of $\mathscr{T}$, such that (a) DPMs of $\mathscr{T}$ are always taken to DPMs, and (b) DPMs related by this transformation are regarded as being empirically equivalent.[17, 18] The claim is, then, that two models of $\mathscr{T}$ should be regarded as representing the same possible world if they are related by a symmetry transformation.[19]

According to this line on symmetries, which Møller-Nielsen dubs the *interpretational* approach (Møller-Nielsen, 2017, p. 1256), in the presence of symmetry-related models, we may first (a) interpret those models as representing the same possible world; then (b) construct a reduced space of KPMs $\tilde{\mathscr{K}}$, each model of which corresponds to a gauge equivalence class of models of $\mathscr{K}$; and *finally* (c) seek to provide a "metaphysically perspicuous characterization" of the models of $\mathscr{K}$ in term of the 'naïve interpretation' of the models of $\tilde{\mathscr{K}}$.[20] This is in contrast

[16] For a concise expression of these points in the language of category theory, see (Weatherall, 2016a,d, 2017). Here, I am eliding the difference between what Dewar calls 'reduction' versus 'internal sophistication'—for the details, see (Dewar, 2019; Martens and Read, 2020; Jacobs, 2021).

[17] For an example of such a symmetry transformation, consider e.g. uniform boosts in Newtonian gravitation (NGT)—this theory being discussed in more detail in §4.1.2. For detailed discussions of this definition of a symmetry transformation, see (Ismael and van Fraassen, 2003; Caulton, 2015; Dasgupta, 2016; Read and Møller-Nielsen, 2020b). Although I concur (obviously!) with the concerns raised by Read and Møller-Nielsen (2020b) regarding such 'epistemic' definitions of symmetries, it is the empirical-equivalence-preserving transformations which will be the *relevant* transformations for my purposes in this book; thus, for simplicity, I adopt here an 'epistemic' approach to symmetries.

[18] As Belot points out, a definition of a symmetry transformation consisting just in condition (a) would be too broad: "Ordinarily, symmetries of theories are hard to come by. But some remarkable theories have atypically large symmetry groups. The definition above effaces this sort of distinction between theories. For if we allow arbitrary permutations of the solutions of a theory to count as symmetries, then the size of a theory's group of symmetries depends only on the size of its space of solutions" (Belot, 2013, p. 322). Though Belot considers more nuanced definitions of symmetry transformations (Belot, 2013, §§ 3–4) (which, for simplicity, are set aside in this work), he argues that all such definitions come into conflict with what Møller-Nielsen calls the 'interpretational approach' to symmetries (discussed below) (Møller-Nielsen, 2017, p. 1256), according to which symmetry-related models represent the same possible world (Belot doesn't use this terminology explicitly in his article, but I take the connection between his discussions and the work of Møller-Nielsen to be transparent). This lends further weight to Belot's claims that the interpretational account should be rejected—see below.

[19] Not all symmetry-related models should be interpreted as corresponding to the same possible world, for consider e.g. the case of Galileo's ship, in which only a subsystem in a model of Newtonian mechanics is boosted. In this case, we have two symmetry-related models, which nevertheless clearly do *not* correspond to the same possible world. In this chapter, I set these complications aside by considering only symmetry transformations which act 'globally' upon the O_i of KPMs of $\mathscr{T}$, rather than upon proper subsystems in those models.

[20] For discussion of what such a "metaphysically perspicuous characterization" consists in, see (Møller-Nielsen, 2017, §3). Note that some advocates of the interpretational approach deny that the seeking of such a characterization is necessary at all—see e.g. (Dewar, 2015, p. 322).

with the *motivational* approach (Møller-Nielsen, 2017, pp. 1256–1257), according to which the existence of symmetry-related models *motivates* us to provide a "metaphysically perspicuous characterisation" of the shared ontology of these models; but tasks (b) and (c) above must be undertaken *before* the interpretative move (a) is made.

In line with Møller-Nielsen, Belot (2018) argues that in several important circumstances, symmetry-related models represent *distinct* physical possibilities. One such case is that of models of GR upon which asymptotic boundary conditions are imposed:[21, 22] (Cf. (Belot, 2018, pp. 964ff.).)

> Consider a sector of GR in which asymptotic boundary conditions have been imposed. Let the space of DPMs be denoted by $\mathscr{D}$, the group of spacetime diffeomorphisms that preserve the asymptotic boundary conditions be denoted by D, and the group of spacetime diffeomorphisms asymptotic to the identity at infinity by D_0. In typical cases of physical interest, we find that D can be viewed as the product of D_0 with a group G that is called the *asymptotic symmetry group* because it can be thought of as acting geometrically at infinity. One then expects to find that G acts also on $\mathscr{D}/D_0$ and that this action is associated in the usual way with the conserved quantities that the imposition of asymptotic boundary conditions bring into existence. In this case, it is natural to take solutions to be gauge equivalent if and only if related by a diffeomorphism in D_0 so that the space $\mathscr{D}/D_0$ can be viewed as representing the physics of the theory without redundancy. But [the interpretational approach] would appear to recommend instead taking the space $\mathscr{D}/D$ to play this role—which would efface the interesting representation of G as the symmetry group of the theory.
>
> (Belot, 2008, p. 201)

(Similar points in this regard are made by Teh (2016) and Dougherty (2020)—in the latter case, it is argued that certain asymptotic gauge transformations do not even preserve *empirical* equivalence.[23, 24, 25]) In light of such cases,[26] there are legitimate reasons to question the interpretational approach, and to take seriously

[21] In this quotation, notation has been altered for consistency with the present work; there is no change in content.

[22] Whether Belot should be classified as endorsing the motivational approach on the basis of this example is, however, unclear, for it appears that in this case Belot does not even consider us *motivated* to find a "metaphysically perspicuous characterisation" of the shared ontology of symmetry-related models. (This point is also made by Luc (2022).)

[23] Note, though, that lack of preservation of empirical content would mean that the transformations under consideration would fail to qualify as symmetries, on an 'epistemic' approach to symmetries.

[24] Of course, there's a fine line between asymptotic conditions (of the kind discussed in these articles) and consideration of subsystem symmetries (of the kind mentioned in footnote 19). In this sense, I cannot *wholly* avoid consideration of the latter in this book.

[25] For some critical engagement with Dougherty's article, see (Gomes and Riello, 2020).

[26] And also—perhaps ultimately more importantly—the issues raised in footnotes 18 and 19.

the motivational alternative.[27] Wherever one stands in this debate, however, the important point to note is that Belot does not subscribe to the former program. Indeed, counting the number of physically inequivalent models which are related by certain symmetries is a crucial aspect of his account of background independence (see §3.6).

[27] I've also argued for this point elsewhere, in (Read and Møller-Nielsen, 2020a,b; Martens and Read, 2020).

3
Classical Background Independence

In this chapter, I discuss the concepts of *general covariance* (§3.1), *diffeomorphism invariance* (§3.2), and *absolute objects* (§3.3). These introduced, in §3.3.3 I assess the possibility of defining background independence as the absence of absolute objects in a theory, before in §3.4 considering definitions of background independence which tie this concept to the absence of either *fixed fields* (§3.4.1), or *absolute fields* (§3.4.2). In §3.5, I consider a definition of background independence in terms of variational principles; in §3.6, I discuss a fifth definition, due to Belot (2011); in §3.7, I consider a recent proposal for background independence from Freidel and Teh (2022), regarding the presence of non-trivial 'corner charges' in a theory. For all of these approaches to background independence, I highlight potential problems and suggest associated revisions; after a brief foray into some general issues in the philosophy of science in §3.8, I close with an integrated discussion in §3.9.

3.1 General Covariance

The first concept in need of discussion is *general covariance*. There's no end to the amount of confusion which this concept has generated over the past century (again, see (Norton, 1993) for the history), but for concreteness and clarity I follow Pooley (2017, p. 115) in giving the following definition:

Definition 1. (General covariance) *A formulation of a theory is* generally covariant *iff the equations expressing its laws are written in a form that holds with respect to all members of a set of coordinate systems that are related by smooth but otherwise arbitrary transformations.*

With such a definition in mind, Friedman states that "the principle of general covariance has no physical content whatever: it specifies no particular physical theory; rather it merely expresses our commitment to a certain style of formulating physical theories" (Friedman, 1983, p. 55). The reason for this—originally pointed out by Kretschmann (1917) in communication with Einstein in 1917—is that *any* theory can be given a generally covariant formulation.[1] Thus, GR does not differ

[1] For further discussion, see (Anderson, 1967; Friedman, 1983; Norton, 1993; Wallace, 2019).

Background Independence in Classical and Quantum Gravity. James Read, Oxford University Press.
 DOI: 10.1093/oso/9780192889119.003.0003

from special relativity (SR), or any other spacetime theory, in virtue of having a generally covariant formulation.

As Pooley points out, however, GR *does* differ from SR in lacking a *non-*covariant formulation (Pooley, 2017, p. 112). This lack of preferred coordinates is due to the fact that the spacetime structures of a generic solution of the theory lack symmetries and so cannot be encoded in special coordinates.[2] This lack of symmetry is entailed by dynamical spacetime structure, but the converse entailment does not hold. Focussing on its possible configurations in a space of DPMs, a *fully* dynamical field, free to vary from DPM to DPM, will in general lack symmetries.[3] The converse is not true, for in principle we can define a theory involving a fixed metric with no isometries, and such a theory will only have a generally covariant formulation (Pooley, 2017, p. 112).[4]

3.2 Diffeomorphism Invariance

3.2.1 Diffeomorphisms

I turn now to the notion of *diffeomorphism invariance*. For a given spacetime theory $\mathscr{T}$ with arbitrary KPM $\mathscr{M} = \langle M, O_1, \ldots, O_n \rangle$, consider a mapping d of the O_i on M, taking $\mathscr{M}$ to a new model $d^*\mathscr{M} := \langle M, d^*O_1, \ldots, d^*O_n \rangle$. If d is smooth, one-one, onto, and has smooth inverse, then it is called a *diffeomorphism*. These diffeomorphisms form a group, denoted $\mathrm{Diff}(M)$.[5]

3.2.2 Two Theories

To facilitate discussion of the notion of diffeomorphism invariance, it is useful to introduce two exemplar spacetime theories. The first—a special relativistic theory

[2] Here, I set aside the possibility of e.g. needing to impose the harmonic coordinate condition in order to actually solve the Einstein equation: see (Stewart, 1991) for further discussion. The situation here is in fact a little delicate: it is perhaps best to say that while the Einstein equation can be solved without harmonic coordinates, these are required in order to show existence and uniqueness (up to diffeomorphisms) of solutions given suitable initial data. (My thanks to Henrique Gomes for discussion on this point.)

[3] Here, I italicize 'fully' in order to emphasize that one can have dynamical spacetimes with symmetries—consider, for example, the subsector of GR in which models are spatially isotropic. In this case, however, one might argue that the symmetries of these spacetimes imply some fixed structure (see e.g. (Lam, 2011) for reasoning along these lines), so the solutions are not to be regarded as being *fully* dynamical. These issues are taken up by Belot (2011), whose article I discuss further below.

[4] Smolin calls into question this claim (Smolin, 2006, p. 201). I endorse Pooley's response, found at (Pooley, 2017, fn. 13). One possibility which Pooley does not consider in making this claim is that of (a) taking a specially relativistic theory set in some Minkowski spacetime, then (b) removing a sufficient number of points to expunge that spacetime of its symmetries, but (c) continuing to use the laws of that theory. Conditional on not modifying the manifold topology by deleting points in this way, Pooley's claim stands.

[5] For details, see e.g. (Wald, 1984, pp. 437ff.).

with KPMs $\langle M, \eta_{ab}, \varphi \rangle$, which, following Pooley (2017, p. 115), I refer to as **SR1**—is a theory of a real scalar field φ, with DPMs picked out by the massless Klein-Gordon equation[6]

$$\eta^{ab} \varphi_{;ab} = 0. \tag{3.2.1}$$

In the KPMs of **SR1**, the Minkowski metric field η_{ab} is *fixed*—i.e. is identically the same in every model.[7] The DPMs are then the proper subset of the KPMs picked out by the requirement that φ satisfy (3.2.1) relative to the η_{ab} common to all the KPMs. Then, (3.2.1) is not an equation for η_{ab} and φ together—it is an equation for φ alone, *given* η_{ab} (Pooley, 2017, p. 115).

Now consider an exemplar general relativistic theory—call it **GR1** (Pooley, 2017, p. 116). Here, KPMs are triples $\langle M, g_{ab}, \varphi \rangle$; in this case, the Lorentzian metric field g_{ab} is no longer fixed, and the DPMs are picked out as a subset of the KPMs by two equations—the field equation (2.1.1), and the analogue of (3.2.1),[8]

$$g^{ab} \varphi_{;ab} = 0. \tag{3.2.2}$$

Here, (3.2.2), unlike (3.2.1), is not an equation for φ *given* g_{ab}. Rather, (3.2.2) and the Einstein equation form a coupled system of equations for g_{ab} and φ *together* (Pooley, 2017, p. 116).[9]

Fixed fields such as the η_{ab} of **SR1** are standardly labeled *background fields* (such a label can also apply to the 'absolute objects' discussed below). In the foundational literature on relativity theory, it is sometimes claimed that the essential novelty of GR is that background fields have been excised from the theory—so that GR qualifies as *background independent* in this sense (Pooley, 2017, pp. 105–107). In §3.2.4, I assess whether this is correct.

3.2.3 Diffeomorphism Invariance

With the above in hand, I am now in a position to discuss diffeomorphism invariance.[10] As Pooley (2017, p. 116) notes, one common such definition—especially in

[6] In **SR1**, a (unique) torsion-free, metric-compatible derivative operator is to be defined in terms of the metric field η_{ab}, and is not another primitive object. η^{ab} is the inverse of η_{ab}.

[7] For some discussion of the formalism underlying the notion of a fixed field, see (Read, 2019a, §A). The crucial point to stress is that, for a fixed field (but not necessarily for e.g. an 'absolute object', which is a concept which I introduce below), the field values at each manifold point are identically the same in every KPM.

[8] g^{ab} is the inverse of g_{ab}.

[9] Discussion of the distinction between theories with fixed fields—such as **SR1**—and theories with no such fixed fields—such as **GR1**—can also be found in e.g. (Giulini, 2007, p. 114) and (Vassallo, 2015). These works also discuss theories akin to Pooley's **SR2** (Pooley, 2017, p. 120), introduced in §3.2.4.

[10] While in this subsection I focus exclusively on diffeomorphism invariance, the setup could in principle be extended to any other symmetries of a given theory.

the post-hole argument foundational literature (see e.g. (Earman, 1989, p. 47))—is the following:

Definition 2. (Diffeomorphism invariance) *A theory $\mathscr{T}$ is* diffeomorphism invariant *iff, if $\langle M, O_1, \ldots, O_n \rangle$ is a DPM of $\mathscr{T}$, then so is $\langle M, d^*O_1, \ldots, d^*O_n \rangle$, for all $d \in \mathrm{Diff}(M)$.*[11]

When applied to **SR1**, Def. 2 yields the verdict that the theory is *not* diffeomorphism invariant. The reason is that Def. 2 requires that if $\langle M, \eta_{ab}, \varphi \rangle$ is a model of the theory, then so is $\langle M, d^*\eta_{ab}, d^*\varphi \rangle$, for an arbitrary diffeomorphism d. However, this definition can only be satisfied when $d^*\eta_{ab} = \eta_{ab}$; for the diffeomorphisms which do *not* satisfy this condition, the model $\langle M, d^*\eta_{ab}, d^*\varphi \rangle$ is not even a KPM of the theory, let alone a DPM (Pooley, 2017, p. 116). What goes for **SR1** here also applies *mutatis mutandis* for any other theory with fixed fields—so, on Def. 2, no such theory can be diffeomorphism invariant.

One might question at this stage, however, whether one should consider unrestricted diffeomorphisms in Def. 2, for a *prima facie* distinct and interesting question is the following: *given* that a diffeomorphism keeps us within the KPMs of the theory, does it also keep us within the DPMs? Following this line, one may choose to restrict in Def. 2 to diffeomorphisms which keep us within the space of KPMs of the theory—and then ask whether such transformations also keep us within the space of DPMs. For example, in **SR1**, on this approach we automatically restrict to diffeomorphisms satisfying $d^*\eta_{ab} = \eta_{ab}$, and then ask whether the transformed model $\langle M, d^*\eta_{ab}, d^*\varphi \rangle = \langle M, \eta_{ab}, d^*\varphi \rangle$ is also a DPM of the theory.

Such an approach, however, does not deliver the correct verdict on the diffeomorphism invariance of theories such as **SR1**. For example, in this theory, the condition $d^*\eta_{ab} = \eta_{ab}$ restricts to diffeomorphisms which act as Lorentz transformations on the geometric objects of the theory. But since the dynamical equation of **SR1** (3.2.1) is Lorentz invariant, it is clear that if $\langle M, \eta_{ab}, \varphi \rangle$ is a DPM of the theory, then so too will be $\langle M, d^*\eta_{ab}, d^*\varphi \rangle$ for d so restricted—delivering the intuitively incorrect verdict that **SR1** *is* diffeomorphism invariant.

Thus, we see that restricting the scope of d in Def. 2 *à la* the above does not deliver a plausible definition of diffeomorphism invariance. One way out here—which captures our desire to assess whether transformations which keep us within the space of KPMs of a theory also keep us within its space of DPMs—is presented by Pooley (2017, p. 116). Letting F_i stand for a fixed field common to

[11] In fact, this is often taken as the definition of general covariance (see e.g. (Friedman, 1983, pp. 51ff.)). As Pooley states, "In arguing for this equivalence, [such authors appear] to overlook the crucial possibility... that a coordinate-free equation relating two geometric objects A and B, can nonetheless be interpreted as an equation for B alone, given a fixed A" (Pooley, 2017, p. 116). This drives a wedge between the notions of general covariance and diffeomorphism invariance, and is discussed further both in this subsection and in §3.2.4 below.

all KPMs (such as η_{ab} in **SR1**), and D_i stand for a dynamical field, Pooley defines diffeomorphism invariance as follows: (Pooley, 2017, p. 117)

Definition 3. (Diffeomorphism invariance, final) *A theory* $\mathscr{T}$ *is* diffeomorphism invariant *iff, if* $\langle M, F_1, \ldots, F_n, D_1, \ldots, D_m \rangle$ *is a DPM of* $\mathscr{T}$*, then so is* $\langle M, F_1, \ldots, F_n, d^* D_1, \ldots, d^* D_m \rangle$*, for all* $d \in \mathrm{Diff}(M)$.[12]

Note that on Def. 3, **SR1** fails to be diffeomorphism invariant, as expected: for an arbitrary diffeomorphism, if $\langle M, \eta_{ab}, \varphi \rangle$ is a solution of **SR1**, then $\langle M, \eta_{ab}, d^* \varphi \rangle$ in general will not be. However, the diffeomorphism invariance of theories with fixed fields is no longer precluded by fiat—rather, it will fail in virtue of the transformed fields no longer satisfying the dynamical equations of the theory in question (thus also avoiding the pitfalls of our above proposed modification of Def. 2). This is a clear advantage of Def. 3. In addition, note that **GR1** is diffeomorphism invariant on unrestricted Def. 2: if $\langle M, g_{ab}, \varphi \rangle$ is a model of the theory, then so is $\langle M, d^* g_{ab}, d^* \varphi \rangle$, for arbitrary d. Since this is the case, **GR1** is also diffeomorphism invariant on restricted Def. 2 (in which we consider only diffeomorphisms which keep us within the KPMs of the theory), and Def. 3.

In the remainder of this work, I fix the sense of 'diffeomorphism invariance' to be that of Def. 3. One important upshot is that general covariance is not identical with diffeomorphism invariance. To illustrate, the dynamical equations of **SR1** are generally covariant; nevertheless, the theory is *not* diffeomorphism invariant. Several authors err in making such an identification—for example, Samaroo writes that "the requirement of general covariance is satisfied to the extent that all of a theory's geometric objects are invariant under $\mathrm{Diff}(M)$; thus, I will refer henceforth to general covariance as diffeomorphism-invariance" (Samaroo, 2011, p. 1075).[13]

3.2.4 Diffeomorphism Invariance and Background Independence

Does diffeomorphism invariance coincide with background independence? Intuitively no, for consider a theory of a real scalar field φ with models $\langle M, g_{ab}, \varphi \rangle$ and dynamical equations

$$g^{ab} \varphi_{;ab} = 0, \tag{3.2.3}$$

$$R^a{}_{bcd} = 0. \tag{3.2.4}$$

[12] This is a special case of the definition of a dynamical symmetry found at (Earman, 1989, p. 45). Cf. footnote 10.

[13] In fact, this statement constitutes a more serious error, for invariance of a theory's geometric objects under diffeomorphisms is surely too strong a criterion for diffeomorphism invariance—rather, the relevant question is whether the transformed fields still satisfy the theory's dynamical equations.

Here, $R^a{}_{bcd}$ is the Riemann curvature tensor associated to the metric field g_{ab}.[14] Once more following Pooley, call this theory **SR2** (Pooley, 2017, p. 120).[15, 16] **SR2** is diffeomorphism invariant on Def. 3. Nevertheless, it is intuitively *not* background independent, for although the metric field is not fixed in all KPMs as with **SR1**, it *is* fixed[17] in all DPMs via (3.2.4). This indicates that diffeomorphism invariance cannot be equated with background independence.

3.3 Absolute Objects

3.3.1 Andersonian Absolute Objects

Anderson (1964, 1967), echoing Kretschmann's observation that any theory can be given a generally covariant formulation, challenges the view that general covariance is the characteristic feature of GR. He claims, rather, that this distinguishing feature consists in GR's representation of its metric affine geometry by a *dynamical object* (Anderson, 1967, p. 83), where such dynamical objects are defined by way of contrast with so-called *absolute objects*. Anderson characterizes absolute objects as follows: *objects the same (up to isomorphism) in all DPMs;*[18] for my purposes, it suffices to elevate this to a definition:

Definition 4. (Absolute object) *A geometric object which is the same (up to isomorphism) in all DPMs of a theory.*

The complement of the class of absolute objects in the class of the geometric objects in the models of a theory is the class of dynamical objects (Anderson, 1967, p. 83). Sometimes, it is claimed (see e.g. (Pitts, 2006)) that a theory is *background independent* just in case it has no absolute objects in its formulation. I assess this proposal in §3.3.3; here I again elevate it to a definition:

Definition 5. (Background independence, absolute objects) *A theory is* background independent *iff it has no absolute objects in its formulation.*

[14] Via the associated (unique) torsion-free, metric-compatible derivative operator—cf. footnote 20.

[15] Note that unlike **SR1**, the metric field of **SR2** is *not* fixed *ab initio*.

[16] One might complain that neither **SR1** nor **SR2** is the most natural formulation of a special relativistic theory of the Klein-Gordon field—rather, KPMs of such a theory should be those constructed by *quotienting* the $\mathscr{K}$ of **SR2** by models the metric fields of which are related by diffeomorphisms. Since, as for Pooley, in this book **SR1** and **SR2** are *toy models* introduced for illustrative purposes, this issue shall here be set aside.

[17] Or, to be more precise, is equivalent up to isomorphism.

[18] For a more formal presentation of Anderson's account, also commenting on the differences between Anderson's original presentation (Anderson, 1967, p. 83) and Friedman's later reformulation (Friedman, 1983, pp. 58–60), see (Pitts, 2006, pp. 348–353). Cf. also (Hiskes, 1984).

3.3.2 Confined Objects

It is worth mentioning one other class of objects which might appear in the models of a theory—referred to by Pitts as *confined objects* (Pitts, 2006, §2) (cf. (Thorne et al., 1973)). Confined objects are structures (not necessarily geometric objects, in the sense of §2.1) which do not change *at all*, under any coordinate transformation (Pitts, 2006, p. 354). As Pitts states, examples include the identity matrix, the Lorentz matrix diag $(-1, 1, 1, 1)$, Dirac γ matrices, Lie group structure constants, and the Kronecker delta and Levi-Civita alternating symbol. Other possible examples of confined objects include the metric signature (Brown, 2005);[19] and the global topology of spacetime (Smolin, 2006; Samaroo, 2011).[20] Smolin implies that all such confined objects should ultimately be excised from our final theories of physics (Smolin, 2006, p. 204)—exploring the extent to which this is both true and possible is an interesting task for future pursuit.[21]

To be completely explicit: fixed fields such as η_{ab} in **SR1** do not qualify as confined objects, because they still have non-trivial associated transformation rules (in the case of η_{ab}, tensorial transformation rules); it is simply that arbitrary diffeomorphisms take η_{ab} to $d^*\eta_{ab}$ and thereby out of the space of KPMs of the theory. By contrast, confined objects, as stated above, do not change at all under any coordinate transformation.

3.3.3 Background Independence and Absolute Objects

As mentioned in §3.3.1, sometimes background independence is equated with the absence of absolute objects in a theory. However, as noted by Pitts (2006), Giulini (2007), Sus (2008, 2010), and Pooley (2017), there are problems with this suggestion:[22]

1. There appear to be cases in which structure that intuitively should count as background is not classified as absolute.

[19] Brown (2005) notes that the metric signature in GR is not a consequence of the field equations. Indeed, in principle (as e.g. Gibbons and Hartle (1990) have discussed), metric signature *change* is possible in GR (or some related theory—for recent discussion in the context of quantum gravity, see (Bojowald and Mielczarek, 2015; Bojowald and Brahma, 2017)). If this work is vindicated, then metric signature may cease to be a confined object.

[20] Mirror symmetry in string theory seems to offer a case in which spacetime topology ceases to be fundamental—for a philosophically oriented introduction, see (Rickles, 2013).

[21] One can read Smolin as demanding a theory whose models share no non-trivial structure—but in that case, one might worry that this can only be true if the theory in question admits of only one model. In fact, this is exactly what Smolin is after—see (Smolin, 2013, §2.1).

[22] This classification of problems is presented in (Pooley, 2017, §7).

2. There appear to be cases in which structure that intuitively should not count as background is classified as absolute.
3. On Anderson's definition, GR itself apparently turns out to have an absolute object (and so should count as background dependent).

I here review each of (1)–(3) in turn.

Case 1: Torretti (1984) gives an example falling into this class, considering a theory of modified Newtonian kinematics in which each model's spatial metric has constant curvature, but different models have different values of that curvature.[23, 24] Because the spatial metric in every model has constant curvature, "such a theory surely has something rather like an absolute object in it" (Pitts, 2006, p. 363). Nevertheless, the failure of the metrics to be locally diffeomorphically equivalent for distinct curvature values entails that the metric tensor does not satisfy Anderson's definition of an absolute object.

What should be made of this example? Arguably, it does not succeed, for, as Pitts observes (Pitts, 2006, pp. 17–18), if one decomposes the spatial metric into a conformal spatial metric density and a scalar density, then the former *is* an absolute object (see (3) below), while the latter, while constant in space and time, is a global degree of freedom (Pitts, 2006, p. 18). Thus, this particular 'counterexample' to the equation of background independence with the absence of absolute objects does not go through.

Clearly, Pitts' response to Torretti's example does not preclude the formulation of other counterexamples falling into this first case.[25] That notwithstanding, a more direct response is also available: simply deny that the relevant structures in these cases do qualify as 'background', on the grounds that they vary between DPMs.[26] To this, Torretti will likely respond that these structures are 'background' in the sense that, in any given model of the theory under consideration, they do not change with respect to a given (temporal) parameter. To this, however, a natural response would be that this constitutes too restrictive a notion of 'background', for essentially all theories—including GR—possess models with 'background structure' in this sense.[27]

Case 2: The best-known example of this type is that of Jones-Geroch 'dust'. Consider a general relativistic theory in which the metric field is coupled to

[23] The kinematics of Newtonian theories will be discussed in more detail in §4.1.

[24] Talk of metrics having curvature can be confusing, since curvature is a property of an affine connection. What's meant here is that the curvature of the derivative operator compatible with the spatial (and temporal) metrics is constant on spacelike hypersurfaces.

[25] Indeed, in §4.3.3, we will see an arguably more successful version of Torretti's counterexample in the context of Kaluza-Klein theory.

[26] Cf. footnote 35, and §3.6.4. Something like this distinction is also drawn by Page (2018).

[27] Consider e.g. a $3 + 1$ decomposition in GR, with constant curvature of the induced metric on the spatial slices.

matter characterized only by a four-velocity field U^a and a mass density. Then, as Friedman states (here, Friedman is thinking in terms of local spacetime physics), "since any two timelike, nowhere-vanishing vector fields defined on a relativistic space-time are d-equivalent, it follows that any such vector field counts as an absolute object... and this is surely counter-intuitive" (Friedman, 1983, p. 59). In response to this example, Pitts, following Friedman, argues that such a 'dust' field violates a ban laid down by Anderson on physically redundant variables (Pitts, 2006, pp. 355–356, 361-362), writing that "the proscription of irrelevant variables serves to eliminate the Jones-Geroch counterexample: the dust four-velocity U^a does not count as an absolute object for GTR + dust because U^a does not exist where there is no dust" (Pitts, 2006, p. 356).[28]

As Pooley states, however, it is unclear whether this response gets to the heart of this counterexample, for "[i]n the context of this theory, the non-vanishing velocity field is, intuitively, as dynamical as the 4-momentum. The trouble arises not because we mistook as indispensable an object that Anderson's definition correctly classifies as absolute. The trouble is that Anderson's definition, intuitively, misclassifies that object" (Pooley, 2017, p. 126). In light of this, Pooley suggests that the notion of absolute objects might not, in fact, be a better candidate than the notion of fixed fields for articulating the sense of 'dynamical' relevant to characterizing background structure, for no such nowhere-vanishing velocity field qualifies as a fixed field, even if it seemingly *may* count as an Andersonian absolute object.[29]

Case 3: A similar strategy might be pursued in the third case, which regards Pitts' claim that GR itself has an absolute object (Pitts, 2006, §8).[30] To see how Pitts reasons to this conclusion, first recall that in so-called *unimodular general relativity* (UGR),[31] since a metric tensor as a geometric object is decomposable into a conformal metric density and a scalar density, the unimodular coordinate condition $\sqrt{-g} = 1$ means that there exists an absolute object in this theory.[32] Once such an example is noticed in the context of UGR, Pitts points out that, in fact, similar reasoning holds for GR itself:

[28] Indices have been modified from coordinate to abstract in this quotation; there is no substantive change in content.

[29] A definition of background independence in terms of the absence of fixed fields is presented and appraised in §3.4.1.

[30] As Pitts notes, this case is also discussed by Giulini (2007); see in addition (Sus, 2008, ch. 3) for further discussion.

[31] That is, the version of GR with cosmological constant—for which the field equations read $G_{ab} + \Lambda g_{ab} = 8\pi T_{ab}$, and the associated Lagrangian is $\mathscr{L} = \mathscr{L}_G + \mathscr{L}_M$, with $\mathscr{L}_G = \int_M d^4x\sqrt{-g}(R - 2\Lambda)$ and $\mathscr{L}_M$ the matter Lagrangian—in which the so-called *trace-free Einstein equation*, $R_{ab} - \frac{1}{4}Rg_{ab} = T_{ab} - \frac{1}{4}Tg_{ab}$, is derived as the extremum of $\mathscr{L}$ via limitation to variations which satisfy the *unimodular coordinate condition* $\sqrt{-g} = 1$ (Earman, 2003, p. 568).

[32] As Pitts mentions (Pitts, 2006, p. 366), there also exists a second version of UGR which admits any coordinates, with the help of a non-variational scalar density; this version of UGR clearly also possesses an absolute object.

> GR has an absolute object! This absolute object [the metric determinant, g] is a scalar density of nonzero weight, because every neighbourhood in every model spacetime admits coordinates (at least locally) in which the component of the scalar density has a value of -1. (Pitts, 2006, p. 366)

The point is that, in any neighborhood of the manifold in any model of GR, one can find a coordinate system such that the object g takes the value -1. Thus, this is a geometric object which is the same (up to isomorphism) in all DPMs of the theory—at least when GR is understood as a local spacetime theory.[33]

As Pooley states, one might accept this verdict without accepting that this automatically means that GR should count as background dependent (Pooley, 2017, pp. 126–127).[34] Such a move is possible if, for example, one rejects the equation of background independence and the absence of absolute objects, and, rather, attempts to return to the view according to which the absence of *fixed fields* (in the sense of **SR1**) is somehow connected with background independence. In that case, for the background independence of GR, one would require that $\sqrt{-g}$ be interpretable as a fixed field—which is not correct. Pooley (2017) explores this option in a further definition of background independence, to which I turn in §3.4.[35]

The upshot of the above three cases is that while some 'counterexamples' to the equation of the background independence of a spacetime theory with its lack of absolute objects do not go through, to the extent that these are successful, the problems they present can be overcome by equating background independence not with the absence of absolute objects, but with the absence of fixed fields. As we have seen in §3.2.4, however, *prima facie* there are problems with this proposal: for example, it cannot account for the background dependence of **SR2**. In the following subsection, I explore how such problems can be overcome.

[33] See (Earman and Norton, 1987, p. 518), and cf. footnote 5.

[34] For an alternative approach, see (Sus, 2008, ch. 3), in which Sus argues for a non-standard definition of *invariance* under a transformation. On this definition, it turns out that GR does *not* have an absolute object.

[35] In connection with UGR, it is worth remarking on some comments made by Earman (2003). In GR with non-zero cosmological constant, Λ is non-dynamical. With this in mind, Earman states that "if Λ is not a universal constant, it too must be varied; and the consequence of this variation of [$\mathscr{L} = \mathscr{L}_G + \mathscr{L}_M$ with $\mathscr{L}_G = \int_M d^4x\sqrt{-g}(R - 2\Lambda)$ and $\mathscr{L}_M$ the matter Lagrangian] is the absurd result that $\sqrt{-g} = 0$, i.e. that the volume of spacetime is zero" (Earman, 2003, p. 562); from this, he infers that Λ must be a universal constant, i.e. constant across all models (KPMs). By contrast, Earman continues, in the UGR formalism, the cosmological constant arises as an integration constant from $\nabla_a(R + T) = 0$, which follows from the UGR field equations and the energy-momentum conservation equation $\nabla^a T_{ab} = 0$, the latter of which in this formalism (and unlike GR) does *not* follow from the field equations, but rather must be postulated separately (Earman, 2003, p. 563). For this reason, in UGR Earman claims that Λ *is* free to vary across models (sometimes, this is claimed to suggest a solution to the infamous *cosmological constant problem*—for skeptical discussion, see (Earman, 2003, §4)). In response to this, one might argue that it is not clear why a non-variational field such as Λ in GR may not differ from model to model (and yet, nevertheless, be non-dynamical in each model)—in fact, we see in §3.6.4 that this possibility may be presented as a problem for Belot's definition of background independence.

3.4 Fixed and Absolute Fields

We have seen that there are problems with the equation of background independence and the absence of absolute objects. Accordingly, in this subsection I turn to two further definitions of background independence. The first, due to Pooley (2017), constitutes a reappraisal of the link between background independence and the absence of *fixed fields* (in the sense of **SR1**); the second is a refinement thereof.

3.4.1 Fixed Fields

Pooley (2017, pp. 122–123, 125), hints at the following definition of background independence:

Definition 6. (Background independence, fixed fields) *A theory is* background independent *iff it has no formulation which features fixed fields.*

The idea here is that while, for example, **SR2** does not feature any fixed fields, it has an alternative formulation—**SR1**—which *does* feature fixed fields; for this reason, **SR2** fails to be background independent. Def. 6 appears to deliver intuitively correct results. To be satisfactory, however, some criterion must be provided for when two *prima facie* distinct theories are, in fact, mere reformulations of the same theory. Drawing upon (Pooley, 2017, p. 122), one might specify this criterion to consist in the following:[36]

Definition 7. (Equivalent theory formulations) *Two theories, $\mathscr{T}_1$ and $\mathscr{T}_2$, are* equivalent formulations *of the same theory when:*

1. *KPMs of $\mathscr{T}_1$ and $\mathscr{T}_2$ involve the same types of geometric object.*
2. *DPMs of $\mathscr{T}_1$ and $\mathscr{T}_2$ are pairwise isomorphic, up to classes of diffeomorphism-related models.*

In Def. 7, (1) can be cashed out as follows: if KPMs of $\mathscr{T}_1$ consist in a certain number of (r, s) tensors, weight w tensor densities, etc., then so too must the KPMs of $\mathscr{T}_2$. Such a criterion is clearly satisfied in the case of **SR1** and **SR2**, in spite of the metric field η_{ab} being a fixed field in the former case.[37] On (2), note that in the case

[36] I discuss theoretical equivalence further in §3.8.

[37] Formulating criterion (1) in this manner excludes many theories related by *dualities* (see §5.3) from qualifying inappropriately as theory reformulations. One legitimate worry about (1) is that it rules as inequivalent theories in which one has trivially repackaged two geometric objects into one, or vice versa. I cannot argue with this objection—in which case, one might incline to one of the more sophisticated definitions of theoretical equivalence discussed in §3.8.1. Nevertheless, I continue to work with this notion of equivalent theory formulations in the remainder of this subsection, for simplicity.

of **SR1** and **SR2**, DPMs of the former theory are isomorphic to DPMs of the latter; it is just that, for each DPM of **SR1**, there is an infinite set of DPMs of **SR2** which differ by a diffeomorphism on the metric field. Nevertheless, as Pooley states, "in the sense that matters, the metric structure of spacetime [in **SR2**] does not differ from DPM to DPM: the g_{ab} in any two DPMs are isomorphic to one another" (Pooley, 2017, p. 122). It is *this fact* which motivates our more lenient formulation of (2) than mere pairwise isomorphism between the respective elements of the $\mathscr{D}$ of $\mathscr{T}_1$ and $\mathscr{T}_2$.

Clearly,[38] with Def. 7 of equivalent theory formulations in hand, Def. 6—of background independence in terms of the absence of fixed fields in said formulations—delivers the correct verdict that both **SR1** and **SR2** are background dependent. One problem with such a definition, however, is that one cannot straightforwardly use it to assess the background independence of a theory *on its own terms*—and absent any argument that there exist *no* formulations of that theory which feature fixed fields, it can be difficult to appraise using Def. 6 whether or not the theory is indeed background independent.[39] Nevertheless, Def. 6 remains valuable; we will see the substantive work to which it can be put in the following chapters.

As a final word on Def. 6, it is worth considering whether this definition does in fact deliver the verdict that GR is background independent. The worry here is that there is a formulation of GR due to Rosen (1940a,b), in which it is proposed to introduce a new fixed metric γ_{ab} at each $p \in M$, the benefit of doing so being that "one imparts tensor character to quantities which in the usual form [of GR] do not have it"—for example, "one can obtain a gravitational energy-momentum density tensor in place of the usual pseudo-tensor" (Rosen, 1940a, p. 147).[40] Since KPMs of the Rosen theory are picked out by tuples $\langle M, g_{ab}, \{\gamma_{ab}\}, \Phi\rangle$, where $\{\gamma_{ab}\}$ denotes the class of γ_{ab} so introduced (one for each point in spacetime), it is, however, clear that this theory will *not* count as a reformulation of GR, on Def. 7; therefore, GR does not violate Def. 6. While Pitts (2016) proposes that fields such as γ_{ab} be replaced simply by a diagonal matrix $\text{diag}(-1,1,1,1)$ (a confined object, in the sense of §3.3.2)—in which case KPMs of the Rosen theory become $\langle M, g_{ab}, \Phi\rangle$—so that in this case the theory *is* a reformulation of GR on Def. 7, since the resulting theory has no fixed fields, there is again here no threat to GR's satisfaction of Def. 6.

[38] And, indeed, arguably by construction—albeit on independently plausible grounds.

[39] Norton stresses that finding theory reformulations of this kind is no easy business (Norton, 1989, p. 829).

[40] My thanks to Brian Pitts for drawing my attention to the Rosen theory. For further recent philosophical discussion on the problem of defining gravitational energy in GR, see e.g. (Hoefer, 2000; Pitts, 2010; Lam, 2011; Curiel, 2019; Read, 2020).

3.4.2 Absolute Fields

Def. 6—of background independence in terms of the absence of fixed fields—appears to deliver correct verdicts on background independence in a range of cases. However, it rests upon a notion of equivalent theory formulation, which, even absent worries regarding this notion being well-defined or appropriately cashed out in Def. 7, leads to the epistemological problem that we might not have access to *all* equivalent formulations of a given theory, the background independence of which we wish to assess. In order to overcome this difficulty, it is useful to first introduce the following notion, of an *absolute field*:

Definition 8. (Absolute field) *A geometric object specified in the KPMs of a theory, which is fixed (up to isomorphism) in all DPMs of that theory.*

Whether a field in the KPMs of a given theory is a fixed field, or is fixed (up to isomorphism) merely in all DPMs via the imposition of a field equation (as for g_{ab} in **SR2**, via (3.2.4)), the field will qualify as an absolute field. Thus, Def. 8 is more permissive than that of a fixed field (in which case the field must be fixed in all KPMs of the theory), but is more restrictive than that of an absolute object (because only geometric objects specified explicitly in the KPMs of a theory can qualify as absolute fields—thus, e.g., the metric determinant might qualify as an absolute *object*, but not as an absolute *field*[41]). This concept of an absolute field in hand, we might proceed to a new definition of background independence:

Definition 9. (Background independence, absolute fields) *A theory is* background independent *iff it has no absolute fields.*

This definition of background independence appears to deliver the correct verdicts that GR is background independent, while both **SR1** and **SR2** are background dependent; moreover, it does so without invoking notions of theory equivalence as with Def. 6. In addition, this definition avoids the difficulty faced by Def. 5 (in terms of absolute objects), that almost *all* spacetime theories of interest feature absolute objects in their formulations, and therefore qualify as background dependent.

[41] In fact, one might worry that Def. 8 is on shaky ground here, for since a metric field can be decomposed as $g_{ab} = \hat{g}_{ab}\sqrt{-g}^{2/n}$, where $\hat{g}_{ab}$ is a conformal metric density; g is the metric determinant; and n is the number of spacetime dimensions (Pitts, 2006, p. 366), we see that, though not *explicit* in the KPMs of e.g. GR, there is a sense in which the metric determinant is straightforwardly still present in those KPMs—and so would seem to qualify as an absolute field. In order for definitions of background independence posed in terms of the absence of fixed fields (see below) to go through, it is therefore clear that much will have to be made of the implicit/explicit distinction. For further discussion of this matter, see §3.8.

All this being said, it is worth reflecting on potential difficulties for Def. 9. Consider, in particular, the following scenario: a given theory $\mathscr{T}$ has KPMs $\langle M, D_1, D_2 \rangle$, where D_1 and D_2 are both dynamical fields (assumed to be of the same tensor rank), with their own associated dynamical equations, and in particular the coupled equation

$$D_1 + D_2 = K, \tag{3.4.1}$$

where K shares the tensor rank of D_1 and D_2, and (crucially) is identical (up to isomorphism) in all DPMs of $\mathscr{T}$. In this case, no field in the KPMs of $\mathscr{T}$ is fixed in all DPMs of this theory. Nevertheless, there *does* exist a plausible absolute field in this theory: K. Thus, Def. 8 appears too weak in restricting to *just* those geometric objects appearing in the KPMs of a given theory; moreover—and potentially more worrisome—Def. 9 arguably delivers an incorrect verdict on the background independence of $\mathscr{T}$, for it does not take into account the existence of the fixed structure K in the DPMs of the theory. While one could widen the scope of Def. 9 to encompass *all* geometric objects fixed in all DPMs of a given theory, this would essentially reduce the definition to Def. 5, in terms of absolute *objects*—bringing with it the associated problems for that definition.

It is plausible that cases such as this do pose a problem for Def. 9. While one might attempt to defend Def. 9 by arguing that scenarios such as that above should be discarded since fields such as K are not *fundamental*, whether one finds this compelling will depend upon one's inclinations in the philosophy of science—see §3.8.1.[42]

3.5 Variational Principles

I turn now to a definition of background independence in terms of variational principles, due to Pooley (2017). In spite of **SR2** and **GR1** sharing the same space of KPMs, there is an important formal difference between these two theories (Pooley, 2017, p. 129). In the case of **GR1**, the space of DPMs picked out by its dynamical equations can be derived from a variational principle via the action

$$S_{\text{GR1}} = \int d^4x \left(\mathscr{L}_G + \mathscr{L}_\varphi \right), \tag{3.5.1}$$

where

$$\mathscr{L}_G = \sqrt{-g}\kappa R \tag{3.5.2}$$

[42] Arguments concerning 'fundamentality' are also key to responses proffered on behalf of advocates of Def. 9 to cases such as that presented in footnote 41. For philosophical discussions of fundamentality, see (Ladyman et al., 2007; Le Bihan, 2018; Crowther, 2019).

is the Einstein-Hilbert 'gravitational' part of the Lagrangian (R being the Ricci scalar and κ a suitable constant), and

$$\mathscr{L}_{\varphi} = \sqrt{-g}g^{ab}\varphi_{;a}\varphi_{;b} \tag{3.5.3}$$

is the 'matter' Lagrangian of the massless Klein-Gordon field. By contrast, the same is *prima facie* not true of **SR2**. This motivates the idea that background independence might be synonymous with all of a theory's dependent variables being subject to Hamilton's principle (Pooley, 2017, p. 130). In turn, this precludes **SR1** from being background independent, since in this case only φ, and not η_{ab}, is subject to Hamilton's principle (in general, this will preclude any theory with fixed fields from being background independent, in line with the morals of §3.3.3). It will also rule out **SR2**—as desired—if the solution space of this theory really is not obtainable from an appropriately formulated action principle.

As noted by both Pitts (2006, p. 357) and Pooley (2017, p. 130), however, it *is* possible to derive the dynamical equations of **SR2**, (3.2.3) and (3.2.4), from an action principle. Rosen (1966) shows that these can be derived from an action in which the $\mathscr{L}_G$ piece in S_{GR1} is replaced with a different term, $\mathscr{L}_S = \sqrt{-g}\Theta^{abcd}R_{abcd}$. This prescription involves a Lagrange multiplier field Θ^{abcd}, in addition to the fields common to **SR2** and **GR1**. In this new action, all the dependent variables are subject to Hamilton's principle.[43] Varying Θ^{abcd} leads to (3.2.4). Since φ does not occur in $\mathscr{L}_S$, varying this field has the same effect as in **GR1**, and leads to the Klein-Gordon equation (3.2.3).[44]

If we want to exclude **SR2** from being background independent on this definition, we need to identify some means of demarcating actions such as that of Rosen as being pathological. To this end, Pooley suggests introducing a further criterion into the variational definition of background independence, to arrive at the following: (Pooley, 2017, p. 130)

Definition 10. (Background independence, variational) *A theory is* background independent *iff its solution space is determined by a generally covariant action, (i) all of whose dependent variables are subject to Hamilton's principle, and (ii) all of whose dependent variables represent physical fields.*[45]

[43] For further discussion of this theory, see (Sorkin, 2002).

[44] One also needs to consider variations of g_{ab}. Rather than the Einstein equation, this leads to an equation that relates Θ^{abcd}, g_{ab} and φ (Sorkin, 2002, p. 696). Such extra equations might legitimately lead one to question whether one is dealing with the same theory at all—indeed, Pooley dubs this theory **SR3** (Pooley, 2017, p. 130).

[45] Versions of GR such as **GR1** seem straightforwardly to satisfy this definition. On the other hand, one might wonder whether e.g. GR in its ADM formulation (see e.g. (Wald, 1984; Gourgoulhon, 2007)) also satisfies Def. 10. Recall that the ADM Lagrangian reads $\mathscr{L}_{\text{ADM}} = N\sqrt{\gamma}\left(\tilde{R} + \left(\gamma^{ik}\gamma^{jl} - \gamma^{ij}\gamma^{kl}\right)K_{ij}K_{kl}\right)$, where N is the *lapse function* (parameterizing the distance between foliations); γ_{ij} is the metric field on foliations; γ is the determinant of γ_{ij}; $\tilde{R}$ is the Ricci scalar associated

The idea is that **SR2** fails to satisfy (ii) because, in Rosen's action, the field Θ^{abcd} is unphysical (as Pooley states, Θ^{abcd} "makes no impact on the evolution of g_{ab} and φ and hence, were it a genuine element of reality, it would be completely unobservable" (Pooley, 2017, p. 130)). Though this seems to overcome the problem in this case, it is unclear whether other *further* cases—such as theories involving 'clock fields'—can be accommodated in a parallel manner (Pooley, 2017, pp. 131–133).[46] To see what is meant here, I turn now to the program of *parameterization*.[47]

Consider a simple case of parameterization, in which one starts with the (Lorentz covariant) action for a theory such as **SR1**, defined with respect to inertial frame coordinates. To construct the parameterized theory corresponding to **SR1**, one treats the four coordinate fields X^{μ} of this formulation as themselves dependent variables ('clock fields'), writes them as functions of arbitrary coordinates, $X^{\mathsf{a}} = X^{\mathsf{a}}(x^{\nu})$,[48] and re-expresses the Lagrangian in terms of these new variables. Hamilton's principle is applied to the original dynamical variables, now understood as functions of x^{ν}, and to the clock fields X^{a}.

As Pitts (2009, pp. 8–9) states, a parameterized special relativistic theory will have clock fields in the Poincaré invariant combination

$$\eta_{\mu\nu} := \eta_{\mathsf{ab}} X^{\mathsf{a}}{}_{,\mu} X^{\mathsf{b}}{}_{,\nu}, \tag{3.5.4}$$

through which the metric field $\eta_{\mu\nu}$ is *defined* in terms of the X^{a}; $\eta_{\mathsf{ab}} = \mathrm{diag}(-1,1,1,1)$ (so that η_{ab} is a *confined object* in the sense of §3.3.2); and the comma denotes partial differentiation with respect to arbitrary coordinates x^{ν}. In the case of **SR1**, stationarity under variations of φ leads to an equation for φ and X^{a} that is satisfied just in case φ satisfies the (Poincaré invariant) Klein-Gordon equation with respect to the X^{a}. Stationarity under variations of the X^{a} yields equations that are satisfied without additional inputs when the first equation is satisfied (Pooley, 2017, pp. 131–132).[49]

In such a theory, do the X^{a} count as 'physical fields'? If they *do* so count in the case of e.g. the parameterized version of **SR1**, then there is a worry that Def. 10

with the induced Levi-Civita derivative operator D_i on foliations; and the *extrinsic curvature* K_{ij} can itself be written $K_{ij} = \frac{1}{2N}\left(\gamma_{ik} D_j \beta^k + \gamma_{jk} D_i \beta^k - \dot{\gamma}_{ij}\right)$, where β^i is the *shift vector*, and dots indicate time derivatives. There exist coordinates $\{x^i\}$ on each foliation Σ_t; each line $x^i = \text{const}$ cutting across foliations defines the *time vector* ∂_t; the shift vector is the projection of the time vector onto each foliation. In this formalism, N and β^i are Lagrange multiplier fields (enforcing the so-called 'Hamiltonian and momentum constraints'); though these are variational fields, one might wonder whether they are *physical*. The straightforward answer appears to be 'yes', since these fields have a clear physical meaning, as delineated above. In that case, ADM GR *does* satisfy Def. 10.

[46] The example of a theory with clock fields is also mentioned by Pitts (2006, p. 359).

[47] Discussed by e.g. Kuchař (1973, 1981).

[48] Having rewritten indices $\mu \rightarrow \mathsf{a}$ ($\mathsf{a} = 0 \ldots 3$) to indicate this elevation in status. The significance of this notational shift shall become clear in §4.2, where a parallel is drawn between clock fields and *tetrad fields*.

[49] In fact, this result holds more generally (Pitts, 2006, p. 9).

delivers the incorrect verdict that such a theory *is* background independent—when intuitively it is *not*, since it is a reformulation of **SR1** (at least in some sense, albeit not necessarily by the standard of equivalent theory formulation introduced above). To resolve this matter, first note with Pooley that, unlike Θ^{abcd} above, the clock fields "certainly encode something physical, since they encode the metrical facts... But there is also a sense in which they do not themselves directly represent something physical: coordinate systems are not physical objects" (Pooley, 2017, p. 132).[50]

While it is true that clock fields, on the above presentation, are *constructed* from a given coordinate system, the relevance of this observation is, ultimately, unclear. The reason for this is that, in the resulting parameterized theory—KPMs of which may be specified to be tuples $\langle M, X^{\mathrm{a}}, \varphi \rangle$—the fields X^{a} manifestly *do* encode physical facts, insofar as they replace the metric field of the original theory. Thus, there are in fact apparently good reasons to take the clock fields to be physical fields, in spite of the means via which they are constructed.

A different question is the following: do the X^{a} qualify as *fixed fields*? Regardless of the theory from which we began (in our example, **SR1**), the answer appears to be *no*, for the X^{a} are subject to a variational principle. However, this *need not be so*, for consider Pitts' observation that, since stationarity under variations of the X^{a} yields equations that are automatically satisfied if equations of motion for the other fields are satisfied, "it does not much matter whether X^{a} are varied or not" (Pitts, 2009, p. 9). If we *do not* vary the X^{a}, then such fields violate Def. 10 of background independence.

The upshot, then, is the following: if the X^{a} *are* varied, then parameterized versions of theories such as **SR1** appear to pose a problem for Def. 10, for such theories, being (in some sense) reformulations of intuitively background dependent theories, are themselves intuitively background dependent. On the other hand, there is room to argue that the X^{a} *should not* be varied—in which case, such parameterized theories *do* come out as background dependent, on Def. 10. To overcome this contingent dependence on whether the clock fields of a parameterized theory are varied or not, one option is to modify (i) of Def. 10 to read: 'all of whose dependent variables *must* be subject to Hamilton's principle, in order to recover the dynamical laws of the theory.'

Finally, I'll change tack, by remarking that Def. 10 should be considered only a *sufficient* condition for background independence, for in principle it is possible to

[50] Note also Pooley's observation that encoding a flat metric via clock fields *à la* (3.5.4) does not uniquely determine those clock fields (Pooley, 2017, p. 132), for if X^{a} corresponds to one such set of fields, then so will any set X'^{a} where the X'^{a} are related to the X^{a} by a Lorentz transformation. This means that the X^{a} contain redundancy; 'internal' Lorentz transformations $X^{\mathrm{a}} \rightarrow X'^{\mathrm{a}}$ should be regarded as mere gauge redescriptions. In my view, the fact that the X^{a} exhibit gauge redundancy does not preclude their being physical fields—or at least, does not preclude some of their degrees of freedom from being counted as physical (cf. e.g. the vector potential of electromagnetism). Thus, the relevance of Pooley's observation here is unclear (cf. §4.2).

construct theories consisting entirely of dynamical fields, with no corresponding action principle (i.e. no Lagrangian associated to the theory).[51] Such theories may qualify, intuitively, as background independent, yet due to the absence of an appropriate action, would not so count on Def. 10.

One illustration of something like this arises through consideration of GR with asymptotic boundary conditions—for example, with the condition that the metric field g_{ab} be flat at spatial infinity (cf. §2.2). There are (at least) three options available to the advocate of Def. 10 in assessing the background independence of this 'theory'. First, they may claim that a theory plus boundary conditions is not itself a new theory, and that background independence of a theory plus boundary conditions (these fixing a $\mathscr{B} \subset \mathscr{K}$) should be assessed with respect to $\mathscr{D} \subset \mathscr{K}$ (recall Fig. 2.1). In this case, since GR comes out as background independent on Def. 10, so too does GR with asymptotic boundary conditions.[52]

A second option here is to consider the specified class $\mathscr{B}$ of BPMs of GR, in this case picked out by the condition of asymptotic flatness, to define the KPMs of a new theory:[53] GR plus these asymptotic boundary conditions. Then, background independence should be assessed with respect to whether the dynamical laws which pick out $\mathscr{D} \cap \mathscr{B} \subset \mathscr{B}$ can be obtained as per Def. 10. Since, however, in this theory we restrict at the level of KPMs to those models satisfying the specified boundary conditions, to derive the laws picking out $\mathscr{D} \cap \mathscr{B}$ from an action principle just is to derive the laws picking out $\mathscr{D}$ from an action principle. On this interpretation—according to which $\mathscr{B}$ picks out the KPMs of a new theory—a theory such as GR with asymptotic boundary conditions can also be background independent, on Def. 10.[54]

A third position is the following. Once again, consider e.g. GR plus asymptotic boundary conditions to be a new theory; KPMs are picked out by $\mathscr{K}$; DPMs are now picked out by $\mathscr{D} \cap \mathscr{B}$—so that in this case boundary conditions are imposed *dynamically*, whereas in the second possibility discussed above, they were imposed at the level of *kinematics*. In this case, in order for the theory to satisfy Def. 10, one would need to derive from an action principle dynamical equations which pick out such a $\mathscr{D} \cap \mathscr{B}$. It is not clear, however, that this can in general be achieved—that

[51] One popular class of examples in modern theoretical physics is that of $D = 6$, $\mathscr{N} = (2,0)$ supersymmetric conformal field theories.

[52] Here is one reason to dispute this option. Suppose that one takes flat asymptotic boundary conditions in the $3+1$ formulation of GR. Since the ADM mass vanishes only for Minkowski spacetime, and the ADM mass is calculated at infinity, as a function of derivatives of the metric, it follows that with sufficiently strong boundary conditions on the metric, a metric that satisfies the Einstein equation has to be flat! And if the metric in every solution under consideration is flat (by virtue of the boundary conditions imposed), it seems incorrect to impute to those solutions properties fundamentally to do with the fact that the flatness of the metric is a *contingent* matter. (My thanks to Henrique Gomes for discussion on this point.)

[53] This possibility was raised in §2.2.

[54] On this approach, such theories in general cease to be diffeomorphism invariant, for a generic diffeomorphism will take one out of the space of DPMs as picked out by $\mathscr{D} \cap \mathscr{B}$.

is, it is not clear that one can in general derive boundary conditions from an action principle. But if this is the case, then it looks like such theories are *not* background independent, on Def. 10 and this third approach—unless, as mentioned above, one regards Def. 10 merely as a *sufficient* condition on background independence.

3.6 Belot's Proposal

In the above, we have seen definitions of background independence in terms of absolute objects (Def. 5); fixed fields (Def. 6); absolute fields (Def. 9); and variational principles (Def. 10)—in all four cases, when one confronts the definitions with concrete cases, quickly their application becomes delicate. In this section, I turn to a fifth account of background independence, due to Belot; the central 'morals' of this approach are claimed to be the following: (Belot, 2011, pp. 2872–2873)

1. Background (in)dependence comes in degrees: some theories are background (in)dependent, others nearly so, and others fall somewhere in between.
2. A theory can fail to be fully background independent in virtue of asymptotic boundary conditions.
3. The extent of the background (in)dependence of a theory is not a strictly formal one: in particular, it depends on how one thinks of the geometric structure of each solution and on what sorts of differences between solutions one takes to be unphysical.[55]

In §§3.6.1–3.6.3, I present how Belot arrives at these results; then, in §3.6.4, I raise some potential problems for Belot's proposal.

3.6.1 Geometrical and Physical Degrees of Freedom

Central to Belot's account are the notions of *geometrical* and *physical* degrees of freedom. Beginning with the former, consider a KPM of a given theory; Belot tells us that we must be able to identify some *geometrical structure* in that model; moreover, there may exist a rule for when the geometrical structures of two KPMs

[55] This interpretative aspect of Belot's definition sets it apart from the definitions presented thus far; indeed, Teitel regards Belot's proposal as "the most promising one formulated to date" (Teitel, 2019, p. 44) in light of this feature. Note that Gryb's proposal for background independence (Gryb, 2010), discussed in §4.4, also has an interpretative dimension.

are *equivalent*. Then, "we can look at the set of spacetime geometries that arise in [models] of the theory and at the set of equivalence classes of such geometries under the relation of geometrical equivalence... We call any set of variables that parameterize this latter set the *geometrical degrees of freedom* of the theory relative to the geometrisation" (Belot, 2011, p. 2874).

To put this slightly differently, Belot's account regarding geometrical degrees of freedom runs as follows. First, for a theory $\mathcal{T}$, take a KPM $\mathcal{M} = \langle M, O_1, \ldots, O_n \rangle$, and identify certain objects in this model as 'geometrical structure'.[56] Marking such structure with a superscript G, we can write this KPM as $\mathcal{M} = \langle M, O^G, O_1 \ldots, O_{n-1} \rangle$. Now consider the geometrical structures O^G and O'^G associated respectively to two models $\mathcal{M}$ and $\mathcal{M}'$ of $\mathcal{T}$. In interpreting the models of $\mathcal{T}$, we may sometimes wish to regard O^G and O'^G as *equivalent* geometrical structure. For example, in models of SR, if $O^G = O'^G = \eta_{ab}$, then clearly these geometrical structures are equivalent, for they are the very same object. Similarly, in GR, we may wish to regard $O^G = g_{ab}$ and $O'^G = d^* g_{ab}$, $d \in \text{Diff}(M)$ as *equivalent* geometrical structures.

Now consider the set $\mathcal{G}$ of the geometrical structures in all DPMs of $\mathcal{T}$. In accordance with the above rule, we can identify subsets of $\mathcal{G}$ which correspond to equivalence classes of geometrical structures. According to Belot's proposal, we should consider the *reduced* set of geometric objects of the theory, call it $\tilde{\mathcal{G}}$, which is constructed from $\mathcal{G}$ by *quotienting* by these classes of equivalent geometries (so that each element of $\tilde{\mathcal{G}}$ corresponds to one such class in $\mathcal{G}$). The degrees of freedom needed to parameterize this latter set $\tilde{\mathcal{G}}$ are what Belot calls the *geometrical degrees of freedom* of $\mathcal{T}$.

With the above in mind, now introduce Belot's notion of physical degrees of freedom. Consider again the space $\mathcal{K}$ of KPMs of a theory $\mathcal{T}$. Often, we wish to interpret two models as representing the same possible world; the models are then said to be gauge equivalent—see §2.2. In the presence of gauge-related models, we may quotient the space $\mathcal{K}$ of $\mathcal{T}$ to arrive at a reduced space $\tilde{\mathcal{K}}$, elements of which correspond to equivalence classes of gauge-related models in $\mathcal{K}$; the thought is that elements of $\tilde{\mathcal{K}}$ contain no redundant 'gauge' degrees of freedom.[57] On Belot's account, the degrees of freedom needed to parameterize $\tilde{\mathcal{D}}$—the gauge-quotiented class of *DPMs* of $\mathcal{T}$—are the theory's *physical degrees of freedom*.[58]

[56] For clarity of exposition, I assume that there is just one such object representing 'geometrical structure' in the model. The presentation here can be extended straightforwardly to the case of multiple geometric objects representing geometrical structure; this, in turn, allows one to accommodate the cases of e.g. Newtonian gravitation theory and Newton-Cartan theory—more on which in Chapter 4.

[57] Again, for simplicity, I am focussing on what Dewar (2019) calls 'reduction' rather than 'internal sophistication'.

[58] The reasons for which physical degrees of freedom are taken to parameterize $\tilde{\mathcal{D}}$ rather than $\tilde{\mathcal{K}}$ are discussed in §3.6.3.

Belot's position on background independence can then be put as follows: a theory is background independent just in case its number of geometrical degrees of freedom coincides with its number of physical degrees of freedom (see Def. 12). The upshot of this definition is that, in a background independent theory, to each model $\tilde{\mathscr{M}} \in \tilde{\mathscr{D}}$ there corresponds *unique* geometrical structure $\tilde{O}^G \in \tilde{\mathscr{G}}$, in the sense of previous subsection. Thus, the geometrical structure $\tilde{O}^G$ acts as a *passport* for the associated model $\tilde{\mathscr{M}}$. In the following subsections, we see how this definition plays out in less abstract terms.

3.6.2 Background Independence

With the above notions of geometrical and physical degrees of freedom in hand, Belot defines four 'degrees' of background independence, as follows: (Belot, 2011, pp. 2877–2878)

Definition 11. (Full background dependence, Belot) *A field theory is* fully background dependent *if it has no geometrical degrees of freedom: every solution is assigned the same spacetime geometry as every other solution.*

Definition 12. (Full background independence, Belot) *A field theory is* fully background independent *if all of its physical degrees of freedom correspond to geometrical degrees of freedom: two solutions correspond to the same physical geometry iff they are gauge equivalent.*

Definition 13. (Near background dependence, Belot) *A field theory is* nearly background dependent *if it has only finitely many geometrical degrees of freedom: quotienting the space of geometries that arise in solutions of the theory by the relation of geometrical equivalence yields a finite-dimensional space.*

Definition 14. (Near background independence, Belot) *A field theory is* nearly background independent *if it has a finite number of non-geometrical degrees of freedom: there is some N such that for any geometry arising in a solution of the theory, the space of gauge equivalence classes of solutions with that geometry is no more than N-dimensional.*

In effect, these definitions measure the degree of background independence by looking at how many of the physical degrees of freedom correspond to geometrical degrees of freedom. The limiting cases are full background dependence (in which there are no geometrical degrees of freedom) and full background independence (in which all physical degrees of freedom correspond to geometrical degrees of freedom); intermediate cases are also possible (Belot, 2011, p. 2878).

3.6.3 Applications

It is instructive to consider the application of the above definitions to certain case studies. Beginning with **SR1**, Belot writes of this theory that "the space of [KPMs] is the space of twice-differentiable φ; and the equation of motion is the Klein-Gordon equation, [(3.2.1)]. Here we have full background-dependence: the Minkowski metric η_{ab} is [*sic*] provides the fixed stage against which the scalar field evolves" (Belot, 2011, p. 2868). The reason for this background dependence is that "each solution has the geometry of Minkowski spacetime, so there are no geometrical degrees of freedom" (Belot, 2011, p. 2878).

On **GR1**, Belot's account also seems to deliver the correct result: "the theory is... fully background-independent: two solutions share the same geometry if and only if they are related by a diffeomorphism if and only if they are gauge equivalent" (Belot, 2011, p. 2878). The expectation is that to each physical degree of freedom there corresponds a distinct geometrical degree of freedom, though whether there is indeed a one-one correspondence here is unclear—I return in §3.9 to assess whether this claim is strictly correct.

Now consider **SR2**. In this theory, the flatness of the (torsion-free derivative operator compatible with the) spacetime metric is not fixed *ab initio* by its being specified as a fixed field; rather, it is determined dynamically by (3.2.4). This means that generic KPMs may be parameterized by non-trivial geometrical degrees of freedom. However, in the *DPMs* of the theory, the metric field g_{ab} is constrained to be flat (i.e. is constrained to have associated compatible torsion-free derivative operator which is flat), and is identical (up to isomorphism) in each model, so that there are no geometrical degrees of freedom.[59] Having made this move, Belot's account again issues the intuitively correct verdict that **SR2** is (fully) background dependent.

Finally, it is worth considering one advertised advantage of Belot's account: that it permits judgments of *intermediate* background independence. To see this, consider the sector of GR in which solutions are asymptotically flat at spatial infinity. Here, $M \cong \mathbb{R}^4$, and the only field is a Lorentzian metric field g_{ab} required to be twice-differentiable, globally hyperbolic, to approximate Euclidean geometry in a suitable sense at spatial infinity,[60] and to satisfy (2.1.1).[61] On this theory, Belot writes:

> The most natural thing to say is that this theory lies between paradigmatic non-background-independent theories like those in which fields propagate against the

[59] This illustrates why, when assessing whether a theory is background independent *à la* Belot, one must do so with respect to DPMs, rather than KPMs.

[60] This fixes $\mathscr{B} \subset \mathscr{K}$ in GR. In this section, however, for simplicity we consider this $\mathscr{B}$ to fix the *KPMs* of the theory of asymptotically flat GR, as per the discussion in §2.1.

[61] This fixes the space of DPMs $\mathscr{D} \subset \mathscr{K}$ of the theory.

backdrop of Minkowski spacetime and paradigmatic background-independent theories like spatially compact general relativity. On the one hand, there are no fields on spacetime, fixed or dynamical, that encode a fixed background structure such as a geometry—indeed, locally the field of the theory has all of the freedom of the metric field of ordinary spatially compact general relativity. On the other, there is also a sense in which the boundary conditions of the theory ensure that any solution has the structure of Minkowski spacetime at infinity—and this is reflected in the fact that the theory is not generally covariant.

(Belot, 2011, p. 2871)

Again, Belot's account offers an intuitively plausible judgment on this example: the theory is nearly background independent, but the fixing of the metric behavior at infinity prevents it from being fully so. To see this, note that in this theory, two solutions correspond to the same geometry if and only if they are related by a diffeomorphism;[62] for Belot, two solutions are gauge equivalent if and only if they are related by a diffeomorphism asymptotic to the identity at spatial infinity.[63] Then, "if one quotients the family of solutions sharing a given geometry by the relation of gauge equivalence, the result is a ten-dimensional space" (Belot, 2011, pp. 2878–2879)—accordingly, the theory satisfies Def. 14.[64]

As an aside, it is interesting to consider what one will make of asymptotically flat GR on Belot's definition if one does not endorse Belot's view that models related by a diffeomorphism asymptotic to the identity at spatial infinity are not to be considered gauge equivalent, but rather embraces the interpretational account, according to which such models are gauge equivalent (recall §2.2). In this case, gauge-equivalent models are those models related by diffeomorphisms, including diffeomorphisms at infinity, so there exists a one-one correspondence between physical and geometrical degrees of freedom, so that the theory comes out as fully background independent.[65]

3.6.4 Problems for Belot's Program

Belot's account appears to deliver the correct verdict on the background independence of a number of examples. Nevertheless, there are some potential bumps

[62] Recall our comment on GR in §3.6.1.

[63] Crucially, Belot's understanding of symmetry-related models as not always being gauge equivalent (see §2.2) is at play here.

[64] Another advertised advantage of Belot's account—as pointed to in §3.6—is that it allows us to make sense of theories which are *ambiguously* background independent (Belot, 2011, §§3.3, 7.1). Note also that Belot's account delivers the correct verdict in other cases, such as that of GR plus Jones-Geroch dust (see §3.3.3).

[65] Cf. §5.3.4.

in the road which should be considered. In this subsection, I first present some methodological concerns, before turning to apparent counterexamples.

The Priority of Geometry

Recall Brown's statement—made on the basis of prior discussions due to Anderson (1967)—that[66]

> Nothing in the form of the equations [of GR] *per se* indicates that g_{ab} is the metric of space-time, rather than a $(0, 2)$ symmetric tensor which is assumed to be non-singular, but, significantly, whose signature is indeterminate.... The 'chronogeometric' or 'chronometric' significance of g_{ab} is not given a priori.
>
> (Brown, 2005, p. 160)

Such a position regarding the geometric import of the fields defined on M in a spacetime theory such as GR is arguably at odds with Belot's approach, which employs a prior specification of geometric objects in any model of the theory in question—with background independence assessed with respect to these 'geometrical' objects. Though this is not a fatal problem for Belot's account, it does highlight that there is more to identifying 'geometry' in spacetime theories than Belot engages with (perhaps, admittedly, deliberately). That said, perhaps any tension here can be ameliorated by arguing that while Brown is concerned with identifying what the spacetime geometry of a theory *is* (a project subsequently taken up by Knox (2019) in her work on spacetime functionalism—see §3.8.2), Belot is concerned with the consequences which follow, once that identification of the geometrical structure in a theory has been made. Indeed, one might quite reasonably retort that the flexibility in Belot's account—built into the approach in order for it to be able to handle cases of ambiguous geometrical commitments in a theory—renders it perfectly capable of being reconciled with the philosophical outlook of Brown.

A second, related point to be made here is this: if one *does* endorse such an Anderson-Brown line, then one may be pushed to provide an account of background independence which treats all fields on a par (such as Def. 6 or Def. 10), rather than one involving any such prior demarcation of geometry. In that case, one might incline away from Belot's approach.

Apparent Counterexamples

Perhaps more significantly, *prima facie* there are worries regarding cases in which Belot's program appears to deliver incorrect verdicts on background

[66] Notation in this quotation has been amended to render it consistent with this book: in particular, I have changed coordinate indices to abstract indices. For the purposes of the point which I wish to make in this subsection the change is not significant.

independence—in part because Belot's definitions (presented in §3.6.2) are (one might argue) insufficiently closely tied to whether the fields of the theory in question are fixed or dynamical; and in part due to the fact that even after addressing this issue, geometrical degrees of freedom (one might argue) *need not* be in one-one correspondence with physical degrees of freedom in an intuitively background independent theory. Let me elaborate in upon each of these two potential problems in turn.

First, consider a spacetime theory with KPMs $\langle M, g_{ab}, \varphi \rangle$, where φ is a real scalar field, subject to the specification that in *every* model, the field g_{ab} adopts an *inequivalent* configuration, and in each case such inequivalent configurations are in one-one correspondence with configurations of the field φ, which are *also* inequivalent across every possible model. In that case, *even though* the g_{ab} field is not subject to dynamical evolution and thereby arguably qualifies as 'background', the physical degrees of freedom of the theory correspond to the geometrical degrees of freedom—so Def. 12 is satisfied.

One possible reply which might be made to this criticism is the following: the statement that the fields in this theory 'are not subject to dynamical evolution' is questionable. If these fields vary between KPMs, then there must be associated dynamical equations for these fields which pick out the DPMs of the theory (recall: one key purpose of dynamical equations is indeed to pick out a subclass of the KPMs as dynamically possible—and this may be the case *even if* the fields of the theory do not change with respect to a to-be-defined parameter *within* any given model). But in that case, the fields of the theory appear to be dynamically variant after all—putting pressure on the initial intuition that such theories possess fixed structure.[67]

Is this response convincing? Arguably, no—for the fact that a certain object varies from model to model, and that one may (at least formally) specify some criterion in order to select the DPMs of the theory in question, does not (one might say) capture the fact that, *within each model*, the piece of designated geometrical structure in play might be dynamically inert: fixed, and not dependent upon the behavior of matter within. In this sense, even if one is capable of constructing 'dynamics' in order to identify the DPMs of the theory in question, this does not overcome the fact that the geometrical structure in such theories is naturally regarded as 'background'. If that is correct, then the above counter maneuvre is not sufficient to save Belot's program on this front.

In any case, consider now a separate worry for Belot's account. For a general relativistic theory in which g_{ab} is coupled to the Φ via (2.1.1) and the dynamical equations for the Φ, there may in principle be many distinct configurations of the Φ for the *same* g_{ab}. To illustrate, consider source-free Einstein-Maxwell

[67] This discussion of course recalls Torretti's point discussed above, which was made in the context of the absolute objects program.

theory (i.e. GR coupled to electromagnetism with no sources), KPMs of which are $\langle M, g_{ab}, F_{ab} \rangle$, where F_{ab} is the Faraday tensor. DPMs of this theory are picked out by the Einstein equation (2.1.1), and the source-free Maxwell equations[68]

$$F^{ab}{}_{;b} = 0, \tag{3.6.1}$$

$$F_{[ab;c]} = 0. \tag{3.6.2}$$

Written in the notation of differential forms, (3.6.1) and (3.6.2) read, respectively

$$dF = 0, \tag{3.6.3}$$

$$d * F = 0, \tag{3.6.4}$$

where $*$ is the Hodge operator for the Lorentzian metric field g_{ab}. Taken together, (3.6.3) and (3.6.4) are formally invariant under

$$F \longleftrightarrow *F. \tag{3.6.5}$$

At the level of electric and magnetic fields,[69] such a transformation can be written

$$E_i \longleftrightarrow -B_i \tag{3.6.6}$$

$$B_i \longleftrightarrow E_i. \tag{3.6.7}$$

This is an example of *electromagnetic duality*.[70] Recalling that the stress-energy tensor of electromagnetism reads

$$T^{ab} = F^{ac}F^{b}{}_{c} - \frac{1}{4}g^{ab}F_{cd}F^{cd}, \tag{3.6.8}$$

one may verify that (3.6.8) is *invariant* under (3.6.5). Thus, if $\langle M, g_{ab}, F_{ab} \rangle$ is a DPM of this theory with stress-energy tensor T^{ab} associated to F_{ab}, then so too is $\langle M, g_{ab}, (*F)_{ab} \rangle$, with inequivalent Faraday tensor, but *identical* stress-energy tensor. Thus, even in GR, there are scenarios in which different configurations of matter fields correspond to the same geometry—thereby violating Belot's definition of background independence. Again, however, it is not clear whether such a result should be considered problematic for Belot—arguably, it is a *merit* of his account that it reveals to us that Einstein-Maxwell theory is (perhaps surprisingly!) *less* background independent than one may initially have thought.

[68] These equations are constructed via the well-known *minimal coupling* scheme—for discussion, see (Read et al., 2018).

[69] Encoded in F_{ab}—see e.g. (Jackson, 1999, p. 556).

[70] For recent philosophical discussion, see (Rickles, 2011; Polchinski, 2017; Weatherall, 2020).

3.7 Non-Trivial Corner Charges

The final account of background independence which I consider in this chapter is one which has been developed in a very recent paper by Freidel and Teh (2022). The definition of background independence proposed by these authors builds upon the discussion of variational principles by Pooley (2017) already presented in §3.5, and is conjunctive in nature. The first conjunct is that the theory under consideration have a solution space determined by a generally covariant action.[71] The second conjunct is more nuanced. In order to explain it, I first need to say a little more on conserved currents in spacetime theories.

In GR, Einstein derived via variational considerations an energy current which, in the modern notation of differential forms, can be written as[72]

$$J_X = C_X + dU_X, \tag{3.7.1}$$

where C_X is the constraint associated with the local (i.e. spacetime-dependent) transformation X, and U_X is the associated 'superpotential'. (Those familiar with Noether's second theorem will recognize (3.7.1) as a generic output when considering theories with local symmetries.) On-shell, $C_X = 0$, in which case, by the standard result that the twice exterior derivative of a form vanishes, it follows that $dJ_X = 0$, which is of course a mathematical identity.[73] Since J_X is on-shell an exact form, one can use Stokes' theorem to write the on-shell charge Q_X as a corner quanity:

$$Q_X = \int_\Sigma dU_X = \int_{\partial\Sigma} U_X, \tag{3.7.2}$$

for a codimension-one hypersurface Σ. So, charges for theories with local symmetries (including GR, which is the particular case under consideration here) are corner quantities.

Freidel and Teh's second conjunctive criterion, then, is that for spacetime transformations (implemented via diffeomorphisms) such corner charges be

[71] This is the same as the first conjunct of Def. 10. Freidel and Teh call this the criterion of 'basic general covariance', and define a theory to be basically generally covariant "just in case diffeomorphisms of its dynamical fields are variational symmetries of the Lagrangian" (Freidel and Teh, 2022, p. 277). This way of putting things is equivalent to the theory having a solution space determined by a generally covariant action, which is how I put this criterion above.

[72] Here, for clarity, I follow the notational conventions of Freidel and Teh (2022).

[73] Famously, on the basis of this result, Klein asserted that conservation laws from Noether's second theorem are physically vacuous, for they are mathematical identities. Einstein demurred, for one only has a mathematical identity on-shell. For a valuable exposition of this episode, see (Brading, 2005); I concur with Brading insofar as I fail to see how an expression being a mathematical identity precludes it from having physical content—in this sense, I side also with Einstein (and, as we will see, with Freidel and Teh).

non-trivial: the thought goes that only in this way do arbitrary diffeomorphisms acquire physical content. In my own words (see (Freidel and Teh, 2022, p. 292) for the original formulation), this second criterion, more formally, is this: for all spacetime diffeomorphisms $d \in \text{Diff}(M)$, the associated corner charge as appearing in (3.7.2) is non-trivial.[74] Putting things together, then, we have this:

Definition 15. (Background independence, corners) *A theory is* background independent *iff (i) its solution space is determined by a generally covariant action, and (ii) for every $d \in \text{Diff}(M)$, the associated corner charge is non-trivial.*

How does this definition play out in practice? As Freidel and Teh (2022) discuss, **SR1** fails Def. 15, for its action is not generally covariant: this is the same outcome as that which we found in the case of Def. 10. More interestingly, in the case of **SR2**, Freidel and Teh demonstrate that the version of this theory derived via variational principles using a Lagrange multiplier field Θ^{abcd} (as presented in §3.5) *does*, in fact, have non-trivial corner charges; thus, the theory qualifies as background independent, by the standards of Def. 15.

Is this the correct verdict? On the one hand, in the preceding sections of this chapter, we proceeded using the intuition that **SR2** is a background dependent theory, for (the claim went) it is simply a reformulation of **SR1**. In a sense, that is correct, when one focuses one's attention on the equations of motions for the two theories *alone*. However, one has to be careful here, because the situation is different when one moves to the action formulation of **SR2** presented in §3.5, for in that case the Lagrange multiplier term in the action encodes *global* degrees of freedom, which in turn are associated with the non-trivial conserved charges.[75] Here is how Freidel and Teh (2022) put the matter:[76]

> [T]he spacetime part has global—but not local—dynamical degrees of freedom which make the solution space for [the theory based upon the action for **SR2**] inequivalent to the theory specified just by the equations of motion $\nabla_a \nabla^b \varphi = 0$ and $R^a{}_{bcd} = 0$. The equation of motion for Θ^{abcd}, in which this dynamical parameter now appears as a potential for the matter field's stress-energy, and the non-trivial corner charge from [the superpotential] U_X, serve to describe how the global spacetime dynamics is [*sic*] coupled to the dynamics of the matter field.
>
> (Freidel and Teh, 2022, p. 293)

[74] It's not totally clear what Freidel and Teh (2022) intend by 'non-trivial': for example, this could be read as 'vanishing', or (more liberally) as 'non-dynamical'. In this book, I'll take the latter reading—this will be relevant in e.g. §4.3.3.

[75] This affords some extra impetus for (with Pooley (2017)) treating the two theories as distinct—cf. footnote 44.

[76] Notation in the following quote has been modified very slightly for consistency with this chapter.

On quantizing, one expects these global, non-perturbative degrees of freedom to show up as instantonic effects.[77] But the main point that I wish to make now, by way of *rapprochement* between Pooley on the one hand and Freidel and Teh on the other, is that while Pooley is *correct* (according to both intuition and Def. 15) that **SR2** is background dependent, he is not correct that the associated dynamically equivalent theory in which Lagrange multipliers are introduced in order to secure said equations by way of varying an action is background independent—at least according to Def. 15. Although Freidel and Teh's definition clashes with intuition in this regard, it has the merit of being able to pick up on non-local physics to which the other definitions of background independence explored in this chapter at least appear to be insensitive.

A mild generalization of Freidel and Teh's remarks leads to the following very straightforward conjecture: if one begins with a theory with fixed fields (such as **SR1**), but then moves to a theory in which the non-dynamical nature of the relevant objects is enforced dynamically (as in the case of **SR2**),[78] the action formulation of the latter theory (if any—this will end up being quite a delicate matter in the case of Newtonian theories, as we'll see in Chapter 4) will in general be background *independent* by the standards of Def. 15, in spite of the background dependence of the starting point in theory space from which one began. In the following chapters, I'll explore this conjecture in more depth in the context of certain specific physical theories.

Before moving on, it is worth making a couple of more critical remarks regarding the Freidel-Teh proposal. First: Freidel and Teh (2022) are completely silent on the physical significance of the charges derived via this machinery in the case of GR—but in that case, there is arguably more to be spelled out regarding the sense in which such charges are 'non-trivial'. Second: in the case of **SR2** *sans* action principle, it is not the case that the theory is *incompatible* with global degrees of freedom—rather, it is simply that the theory is *silent* on such degrees of freedom.[79] This, clearly, is conceptually distinct from fixing or dynamically constraining those degrees of freedom. But in that case, it's not obvious that Freidel and Teh's implication for their verdict that the action formulation of **SR2** is background independent, but the formulation of the theory directly in terms of equations of motion is background dependent, is warranted.[80]

Let me close this section with one further—quite straightforward—remark. Just as in the case of Def. 10, the entire Freidel-Teh approach is conditional on the theories under consideration admitting of formulations in terms of variational

[77] I'll turn in earnest to the matter of quantization in §5.1.

[78] Recall again from footnote 39 that in general making these moves is far from trivial or guaranteed.

[79] For related discussion, see (Wolf and Read, 2023).

[80] My thanks to Henrique Gomes for discussion on these points.

principles—so, once again, one might argue that it can supply, at best, a *sufficient* condition for background independence.

3.8 Functionalism and Structuralism

In this final subsection of the chapter, I discuss some aspects of *structuralist* and *functionalist* approaches to the ontology of spacetime theories—particularly in connection with recent work of Knox (2011, 2013, 2014, 2019). This discussion will be of value when it comes to assessing the background independence of various spacetime theories in the following chapters.

3.8.1 Structuralism

Following Weatherall (2016a,d, 2017), we can understand two theories $\mathscr{T}_1$ and $\mathscr{T}_2$ be *theoretically equivalent* just in case (i) there exists a (potentially many-many) mapping from solutions of $\mathscr{T}_1$ to solutions of $\mathscr{T}_2$, and (ii) empirical content is preserved by this map.[81, 82, 83] Consider two theoretically equivalent theories $\mathscr{T}_1$ and $\mathscr{T}_2$, with respective spaces of KPMs $\mathscr{K}_1$ and $\mathscr{K}_2$. Elements of $\mathscr{K}_1$ can be written $\langle M, O^1_1, \ldots, O^1_m \rangle$; elements of $\mathscr{K}_2$ can be written $\langle N, O^2_1, \ldots, O^2_n \rangle$. Assuming that the O^1_i differ from the O^2_j in some way (perhaps they are tensor densities of differing weight, perhaps they are tensor fields of different rank, or perhaps they are tensor densities of the same rank, which obey different dynamical equations), on a *naïve realist* interpretation of $\mathscr{T}_1$ and $\mathscr{T}_2$, models of these theories related by the theoretical equivalence map should in general be regarded as

[81] The distinction between Def. 7 and this definition of theoretical equivalence is worthy of discussion; the former is both stronger and weaker than the latter. It is stronger for two reasons. First, it involves the condition that the KPMs of two theories which are to be regarded as formulations of one another be specified in terms of the same types of geometric object; the latter definition does not involve this requirement. Second, Def. 7 is stronger than this notion of theoretical equivalence, because it requires that the DPMs of the two theories in question be isomorphic—but this is not a condition in this notion of theoretical equivalence. On the other hand, Def. 7 is weaker, for it is a purely *formal* notion—no condition of empirical equivalence is part of this definition.

[82] The approach to theoretical equivalence presented by Weatherall (2016a,d, 2017) is couched in the language of category theory; for the sake of simplicity, I seek here to eschew category theory. For more general background on theoretical equivalence, see (Weatherall, 2019a,b); for further justification for thinking of theoretical equivalence in terms of category theory (and the associated notion of categorical equivalence), see (Weatherall, 2021).

[83] Establishing the empirical content of (a model of) a theory is, of course, no easy business. Van Fraassen (1980) speaks famously of the 'empirical substructures' of models; absent further details, however, this leaves open the extent to which the empirical content of a theory is a function of (say) other theoretical commitments, or of e.g. pragmatic factors (for some further recent discussions of these points, all from somewhat different angles, see (Dasgupta, 2016; Read and Møller-Nielsen, 2020b; Wallace, 2022b)). For my purposes in this subsection, I simply assume that a tolerably unambiguous notion of the empirical content of (a model of) a theory is to be had.

representing distinct possible worlds. By contrast, according to what I will call here a *structuralist* interpretation of $\mathscr{T}_1$ and $\mathscr{T}_2$, models of these theories related by the theoretical equivalence map may ultimately be regarded as representing the *same* possible world.[84] One corollary of this position is that a structuralist will feel free to move between representing the world under consideration with the element of $\mathscr{K}_1$ versus the associated element of $\mathscr{K}_2$—for there is no substantive difference of content involved in this choice.[85]

This distinction between naïve realist and structuralist approaches to the interpretation of models of a given theory is of value when considering Def. 9 of background independence, in terms of absolute fields. In §3.4.2, I presented two theories: the first ($\mathscr{T}_1$) with KPMs $\langle M, K\rangle$, where K is an absolute field; the second ($\mathscr{T}_2$) with KPMs $\langle M, D_1, D_2\rangle$, where D_1 and D_2 are not identically the same in all DPMs of the theory, but *are* subject to the constraint (3.4.1). To recall, the worry for Def. 9 was that, whereas $\mathscr{T}_1$ violates Def. 9, $\mathscr{T}_2$ does not, for it has no absolute fields in its KPMs—but this, arguably, is incorrect, for there is a sense in which $\mathscr{T}_2$ *still* has an absolute field in its formulation (namely, K), even if this is not explicitly presented in the KPMs of the theory.

On both the naïve realist and structuralist approaches, there are problems with Def. 9. On the former, $\mathscr{T}_1$ is to be regarded as background dependent, whereas $\mathscr{T}_2$ is to be regarded as background independent. As mentioned, however, arguably this is incorrect, for there is a sense in which fixed structure does exist in $\mathscr{T}_2$. That said, it is possible that the naïve realist advocate of Def. 9 can assert here that this is not problematic—for what might be argued to be at issue is whether the *fundamental* fields in the KPMs of a given theory constitute 'background' or not. The worry about this, however, is that it places an implausible amount of weight on what one might take to be no more than presentational choices regarding the geometric objects appearing in the KPMs of the theory under consideration. On the other hand, for the structuralist, there is no difference in the representational content of the models of these two theories—so the choice between $\langle M, K\rangle$ versus $\langle M, D_1, D_2\rangle$ is a merely conventional one. In that case, there is a sense in which there *also* exist absolute fields in $\mathscr{T}_2$, causing this theory to violate Def. 9. In this latter case, however, the worry—as already discussed in §3.4.2—is that Def. 9

[84] On the interpretational approach to theoretically equivalent theories, this claim regarding the physical equivalence of the two theories can be made *ab initio*; on the motivational approach, the claim can only be made once a perspicuous understanding of the common ontology of the two models in question is attained (Read and Møller-Nielsen, 2020a). Just as in the case of symmetries, two different ways of articulating the common ontology of these models related by the theoretical equivalence mapping are what Dewar calls 'reduction' and 'internal sophistication'—see (Dewar, 2019; Martens and Read, 2020; Jacobs, 2021).

[85] Of course, one of the two models might be preferred on the grounds of presentational perspicuity, etc. The structuralist might wish to regard the choice between using models of $\mathscr{T}_1$ versus models of $\mathscr{T}_2$ as a conventional one—for recent discussion of conventionalism in the foundations of physics, see (Dürr and Read, 2023).

simply collapses to Def. 5, in terms of absolute objects. The upshot, then, is that Def. 9 is only unproblematic if one is a naïve realist about the interpretation of physical theories, and one regards the geometric objects specified explicitly in the KPMs of a given theory as being ontologically privileged.

3.8.2 Spacetime Functionalism

Having presented this difference between naïve realism and structuralism, it will also prove to be of value to say something at this point on the *spacetime functionalist* program of Knox (2011, 2013, 2014, 2019), which grew out of the work on spacetime theories by Brown (2005) (cf. §3.6.4). In brief, Knox's program can be summarized as follows: "I propose that the spacetime role is played by whatever defines a structure of local inertial frames" (Knox, 2019, p. 122). That is, for Knox, in any given theory, the degrees of freedom in the models of that theory which are to be regarded as 'spatiotemporal' *just are* those which define a structure of local inertial frames.[86] This proposal in mind, suppose now that we have two theories which are theoretically equivalent in the sense of the previous section.[87] Of such a case, Knox writes:

> I argue that these theories do not, in fact, represent cases of worrying underdetermination. On close examination, the alternative formulations are best interpreted as postulating the same spacetime ontology. In accepting this, we see that the ontological commitments of these theories cannot be directly deduced from their mathematical form. (Knox, 2011, p. 264)

In making claims such as the above, Knox moves from the identification of structures in models of the two theories in question both qualifying as spatiotemporal in light of her functionalist criterion, to a structuralist claim that those structures should be interpreted as representing the *very same* ontology.

Such a move is not innocuous: to make clear the sense in which it is not, it is useful to follow the philosophy of mind literature in distinguishing *role functionalism* from *realizer functionalism* (McLaughlin, 2006; Levin, 2018). While realizer functionalists "take a functional theory merely to provide definite descriptions of whichever lower-level properties satisfy the functional characterizations" (Levin, 2018, §3.4), role functionalists advance an ontological thesis according to which the structure under consideration is *identified* with whatever plays the functional

[86] For details here, including a clear characterization of what is meant by a 'local inertial frame', see (Knox, 2013, 2019); for critical discussion of Knox's proposal, see (Baker, 2021; Read and Menon, 2021).

[87] We see this in the cases of Newtonian gravitation and Newton-Cartan theory, and GR and teleparallel gravity, to be discussed later in this chapter. Knox claims that there are differences between these two cases—for a response, see (Read and Teh, 2018, 2022).

role. Transplanting this distinction onto the spacetime case, it is clear that Knox's spacetime functionalism, combined with advocation of a brand of structuralism akin to that presented in the previous section, commits Knox to a version of role functionalism: the spacetime ontology (rather than merely description) of any given theory *just is* that structure which picks out the local inertial frames: hence, Knox's spacetime functionalist project is more than a merely descriptive thesis of the realizer functionalism kind.[88]

In this book, I seek—in the spirit of naïve realism—to assess the background independence of spacetime theories (in the first instance) *on their own terms*, without reference to other theories to which they may be theoretically equivalent. This is not intended as a critique of either structuralism or functionalism *à la* Knox, but rather in the spirit of maximal generality: by assessing the background independence of theories in such a manner, one can then choose to remain a naïve realist and take those assessments to apply to the theories under consideration *tout court*; on the other hand, if one is a structuralist, one may ultimately choose to appraise the background independence of a theory by reference to other theories to which that first theory is theoretically equivalent—where the ontological pictures of one of these theories might be preferred on Knoxian functionalist grounds, or on the grounds of eliminating gauge degrees of freedom, etc.[89] When I speak in favor of this approach at later points in this work, I am motivated principally by these pragmatic considerations.

3.9 Close

In this chapter, I discussed *general covariance* (§3.1), *diffeomorphism invariance* (§3.2), and *absolute objects* (§3.3). Such background in hand, I then presented six definitions of background independence:

- Def. 5, in terms of the absence of absolute objects in a theory (§3.3.1).
- Def. 6, in terms of the absence of a theory formulation involving fixed fields (§3.4.1).
- Def. 9, in terms of the absence of absolute fields in a theory (§3.4.2).
- Def. 10, in terms of variational principles (§3.5).
- Def. 12, in terms of matching geometrical and physical degrees of freedom (§3.6.2).
- Def. 15, in terms of the presence of non-trivial corner charges in a theory (§3.7).

[88] See (Duerr, 2021) for related discussion. I'm grateful to Ruward Mulder for insights on these issues.
[89] These projects are not unrelated: see (Knox, 2014).

All six of these definitions face various concerns. In the following chapters, I embark upon an in-depth application of the definitions to various classical spacetime theories. My purpose in doing so is not only to provide a more concrete appraisal of the background independence of these theories than has hitherto been achieved, but also to continue to assess the relative merits of the above definitions, and the extent to which any such worries can be overcome.

I want to close this chapter with two final comments. First, some readers might be perplexed as to why I appear not to have engaged more with certain other well-known and respected articles on background independence—in particular, those by Rovelli (1997), Giulini (2007), and Rickles (2008b). The reason for this is that all such articles have been thoroughly assessed and scrutinized by Pooley (2017); ultimately, Pooley incorporates them into his own analyses, with which I have engaged substantially throughout this chapter. To say more on those articles would simply be to parody Pooley's own excellent work—so, I refer readers to his paper for further engagement with those articles.

Second, and more substantially: up to this point, I have not mentioned what's known as the *action-reaction principle*. Roughly speaking, this principle states that every (geometric) object in a theory must be dynamically coupled to other bodies in the theory, such that it affects the motions of those bodies ('action'), and vice versa ('reaction'). This principle is taken very seriously by philosophers such as Brown (2005), but it is important to stress that absent further details it remains vague; putting further meat on the bones can lead the principle to amount to (for example) one of the approaches to background independence already considered in this chapter. To illustrate, Anderson and Gautreau identify explicitly the action-reaction principle as a key motivation for the 'no absolute objects' agenda, writing: "Roughly speaking, an absolute object affects the behaviour of other objects but is not affected by these objects in turn" (Anderson and Gautreau, 1969, p. 185, fn. 6).[90]

Another reason for my not having engaged further with the action-reaction principle in this chapter is that said principle has become entangled in the literature with high-level philosophical and metaphysical debates about which I wish to remain neutral (at least for the purposes of this project). For example, whether one thinks that the action-reaction principle is satisfied or violated by the Minkowski metric of special relativity, or the structures of Newtonian spacetimes (on which, see Chapter 4), will depend upon whether one is a substantivalist or a relationist about such an object: on the former view, such objects straightforwardly seem to act (on material fields) but not to react; on the latter view, it's at least less obvious

[90] For further discussion, see (Norton, 1993, p. 844). In the particular context of the absolute objects program, it's in fact not obvious that non-absolute objects are invariably action-reaction principle satisfying objects—this point is made by Pooley, who stresses that an object "might affect without being affected and yet have non-trivial dynamics of its own" (Pooley, 2017, p. 123).

that this is the case.[91] For example, Brown (2005) espouses a particular form of relationism about the above-mentioned geometrical structures (associated with his 'dynamical approach' to spacetime theories—see (Brown and Read, 2021) for a recent review); it is therefore not so surprising that, in an article with Lehmkuhl on the action-reaction principle, he writes that

> [w]hat is debatable is not the claim that GR is consistent with the action-reaction principle, it is the claim that the older theories involving absolute space-time structures are not. More specifically, it is the assertion that such structures act in the relevant sense. Clearly, if they do not..., then the fact that material bodies do not act back on them constitutes no violation of the action-reaction principle.
> (Brown and Lehmkuhl, 2015, p. 2)

Now, it's one thing for an approach to background independence (e.g. that of Belot (2011)) to invoke interpretative issues in a precise and well-controlled manner. But what the above illustrates is that satisfaction of the action-reaction principle is rather more nebulous than that, and seems to track one's more general background philosophical commitments (here, on the substantivalism-relationism debate). Given this lack of precise control over the notion, I conclude that the action-reaction principle—conceptually and historically interesting and important as it is—is best set aside in these discussions.

[91] Partly, the answer which one gives here will depend upon whether a relationist counts themselves as an *eliminativist* about said structure, or rather as a *reductionist*. If the former, it's hard to see how there could be violations of the action-reaction principle—for the object under consideration does not really exist! If the latter, such violations may be possible, for on reductionist relationism such geometrical structures do exist—it is simply that they are not to be regarded as being fundamental. In that case, the possibility of violations of the action-reaction principle seems more plausible.

4
Classical Theories of Spacetime

In this chapter, I evaluate five classical spacetime theories—Newtonian gravitation theory (§4.1.2), Newton-Cartan theory (§4.1.3),[1] teleparallel gravity (§4.2.2), Kaluza-Klein theory (§4.3), and shape dynamics (§4.4)[2]—with respect to the question of their background independence. In each case, I present a comprehensive application and discussion of the definitions of background independence presented in Chapter 3; many important points regarding both these definitions, and the theories to which they are applied, emerge from this analysis.

4.1 Newtonian Gravity and Newton-Cartan Theory

In the field formulation of NGT, gravity is mediated by a potential; given a (non-constant) change of gravitational potential, the matter distribution will change.[3] NGT is theoretically equivalent to another theory,[4] known as *Newton-Cartan theory* (NCT). In NCT, by contrast with NGT, gravity is 'geometrized' in a manner similar to that of GR: the geometrical properties of spacetime depend on the distribution of matter, and, conversely, gravitational effects are manifestations of this geometry.[5]

In this section, I first introduce in §4.1.1 the formal notion of a *classical spacetime*, this being the setting for both NGT and NCT. These theories I then discuss in §§4.1.2–4.1.4.[6] Finally, in §4.1.5, I engage in an in-depth assessment of the background independence of both of these theories.

[1] In fact, two different versions of Newton-Cartan theory.

[2] In fact, both scale-invariant Newtonian particle dynamics and relativistic shape dynamics 'proper'.

[3] See §4.1.2. In fact, the precise sense in which this is so is underdetermined in NGT—see footnote 15.

[4] In the sense of the second notion of theoretical equivalence discussed in the previous section, i.e. that there exists an isomorphism between $\tilde{\mathscr{K}}_1$ of $\mathscr{T}_1$ and $\tilde{\mathscr{K}}_2$ of $\mathscr{T}_2$ such that the empirical content of (the elements of) $\tilde{\mathscr{D}}_1 \subset \tilde{\mathscr{K}}_1$ is identical to the empirical content of (the associated elements of) $\tilde{\mathscr{D}}_2 \subset \tilde{\mathscr{K}}_2$.

[5] One might question whether Einstein would speak about the situation in this way, for similar reasons to those underlying his objection to the claim that GR 'geometrizes' gravity—see (Lehmkuhl, 2014).

[6] Much of my presentation of the formal aspects of these theories will follow the classic texts (Friedman, 1983; Earman, 1989; Malament, 2012). In particular, I don't claim any originality whatsoever for the theorems to follow, and indeed most are based upon the beautiful presentation of Malament (2012).

Background Independence in Classical and Quantum Gravity. James Read, Oxford University Press.
 DOI: 10.1093/oso/9780192889119.003.0004

4.1.1 Classical Spacetimes

Define a *classical spacetime* to be a quadruple $\langle M, t_{ab}, h^{ab}, \nabla \rangle$, where M is a smooth four-dimensional differentiable manifold; t_{ab} is a smooth, symmetric tensor field on M of signature $(1,0,0,0)$; h^{ab} is a smooth, symmetric tensor field on M of signature $(0,1,1,1)$; ∇ is a derivative operator on M;[7] and the following three conditions hold:

$$h^{ab} t_{ab} = 0, \tag{4.1.1}$$

$$\nabla_a t_{bc} = 0, \tag{4.1.2}$$

$$\nabla_a h^{bc} = 0. \tag{4.1.3}$$

I refer to (4.1.1) as an *orthogonality* condition, and (4.1.2) and (4.1.3) as *compatibility* conditions. For the definition of the 'signature' of degenerate objects such as t_{ab} and h^{ab}, see (Malament, 2012, pp. 249–250). Given its degeneracy, h^{ab} does not have a unique inverse.

Let us consider now the interpretation of the t_{ab}, h^{ab} and ∇ in a classical spacetime. Beginning with t_{ab}, this can be considered a 'temporal metric'—though, as we have seen, strictly speaking it does not satisfy the metric non-degeneracy condition. Given any vector θ^a at a point $p \in M$, we can take its 'temporal length' to be $\left(t_{ab}\theta^a\theta^b\right)^{1/2}$. We can classify θ^a as either *timelike* or *spacelike*, depending on whether its temporal length is positive or zero, respectively. It follows from the signature of t_{ab} that the subspace of spacelike vectors at any point is three-dimensional (Malament, 2012, p. 250). At any $p \in M$, we can find a covector t_a, unique up to sign, such that $t_{ab} = t_a t_b$ (Malament, 2012, pp. 250–251). We say that a classical spacetime is *temporally orientable* if there exists a continuous, *globally defined* covector field t_a that satisfies this condition at every $p \in M$. Any such field t_a is called a *temporal orientation.* A timelike vector θ^a is *future-directed* relative to t_a if $t_a\theta^a > 0$; otherwise it is *past-directed* (Malament, 2012, p. 251).[8]

In the following, I restrict to temporally orientable classical spacetimes. Since in this case t_{ab} can be constructed from t_a, I accordingly denote classical spacetimes which satisfy the temporal orientability condition by $\langle M, t_a, h^{ab}, \nabla \rangle$.[9] For such spacetimes, (4.1.1) and (4.1.2) can be written, respectively,

[7] For the formal definition of such a derivative operator, see e.g. (Malament, 2012, p. 49).

[8] We understand a smooth curve to be *timelike* (respectively *spacelike*) if its tangent vectors are of this character at every point along the curve. Similarly, a timelike curve is understood to be *future-directed* (respectively *past-directed*) if its tangent vectors are so at every point along the curve (Malament, 2012, p. 251).

[9] If a classical spacetime admits one temporal orientation t_a, then it also admits a second temporal orientation, $-t_a$ (Malament, 2012, p. 251). In cases in which I restrict attention to classical spacetimes that are temporally orientable, I by convention select the temporal orientation with positive sign.

$$h^{ab}t_b = 0, \tag{4.1.4}$$

$$\nabla_a t_b = 0. \tag{4.1.5}$$

From (4.1.5), it follows that t_a is a closed form. Hence by the Poincaré lemma at least locally it must be exact, i.e. of the form $t_a = \nabla_a t$, for some smooth function t. We call t a *time function* (Malament, 2012, p. 251). Call a hypersurface $S \subset M$ *spacelike* if, at all $p \in S$, all vectors tangent to S are spacelike; this is equivalent to the constancy of time functions on S. We can think of spacelike hypersurfaces as 'simultaneity slices'. If M is simply connected, then there exists a globally defined time function (Malament, 2012, p. 252). In that case, spacetime can be decomposed into a one-parameter family of global simultaneity slices.

Now consider h^{ab}. Informally, this can be considered a 'spatial metric', though the details of how h^{ab} assigns lengths to vectors are subtle. We cannot take the spatial length of a vector θ^a to be $\left(h_{ab}\theta^a\theta^b\right)^{1/2}$, because the latter is not well-defined. If, however, θ^a is *spacelike*, then we can use h^{ab} to assign a spatial length to it indirectly—for the details as to how, see (Malament, 2012, p. 253). Finally, note that because h^{ab} is not invertible, we cannot raise *and* lower indices with it. We can, however, *raise* indices: for example, if $R^a{}_{bcd}$ is the Riemann curvature tensor associated with ∇ (see below), then we understand $R^{ab}{}_{cd}$ to be the field $h^{be}R^a{}_{ecd}$ (Malament, 2012, p. 256).

Now consider the derivative operator ∇ in a classical spacetime $\langle M, t_a, h^{ab}, \nabla \rangle$. I first present the following general result regarding derivative operators on M: (Malament, 2012, p. 51)

Proposition 1. *Let ∇ and ∇' be derivative operators on M. Then there exists a smooth symmetric tensor field $C^a{}_{bc}$ on M satisfying, for all smooth tensor fields $\alpha^{a_1 \ldots a_r}_{b_1 \ldots b_s}$ on M:*

$$\left(\nabla'_c - \nabla_c\right)\alpha^{a_1 \ldots a_r}_{b_1 \ldots b_s} = \alpha^{a_1 \ldots a_r}_{d b_2 \ldots b_s} C^d{}_{cb_1} + \ldots + \alpha^{a_1 \ldots a_r}_{b_1 \ldots b_{s-1} d} C^d{}_{cb_s} - \alpha^{d a_2 \ldots a_r}_{b_1 \ldots b_s} C^{a_1}{}_{cd} \tag{4.1.6}$$

$$- \ldots - \alpha^{a_1 \ldots a_{r-1} d}_{b_1 \ldots b_s} C^{a_r}{}_{cd}$$

Conversely, given any derivative operator ∇ on M and any smooth symmetric tensor field $C^a{}_{bc}$ on M, if ∇' is defined by (4.1.6)*, then ∇' is also a derivative operator on M.*

Proof: See (Malament, 2012, pp. 51–52). □

In what follows, if ∇' and ∇ are derivative operators on M that, together with the field $C^b{}_{cd}$, satisfy (4.1.6), then I shall, following Malament (2012, p. 53), write $\nabla' = \left(\nabla, C^b{}_{cd}\right)$. $C^b{}_{cd}$ is sometimes called the *difference tensor* of ∇' and ∇. Now, for a classical spacetime, the conditions (4.1.2) and (4.1.5) do not determine a unique ∇. In fact, we have the following result: (Malament, 2012, p. 256)

Proposition 2. *Let $\langle M, t_a, h^{ab}, \nabla \rangle$ be a classical spacetime. Let $\nabla' = \left(\nabla, C^b{}_{cd}\right)$ be a new derivative operator on M. Then ∇' is compatible with t_a and h^{ab} iff $C^a{}_{bc}$ is of the form*

$$C^a{}_{bc} = 2h^{ad} t_{(b} \kappa_{c)d}, \tag{4.1.7}$$

where κ_{ab} is a smooth, antisymmetric tensor field on M.

Proof: See (Malament, 2012, pp. 256–257). □

Sometimes, κ_{ab} is called the *Newton-Coriolis 2-form* (a term, to my knowledge, introduced by Geracie et al. (2015)).

It is also worth considering some properties of the Riemann tensor $R^a{}_{bcd}$ associated to ∇ in a classical spacetime. As always (see e.g. (Malament, 2012, p. 69)), $R^a{}_{bcd}$ satisfies the algebraic relations $R^a{}_{b(cd)} = R^a{}_{[bcd]} = 0$. The conditions (4.1.2) and (4.1.5) further imply that $t_a R^a{}_{bcd} = R^{(ab)}{}_{cd} = 0$. It follows that if we raise all indices with h^{ab}, the resulting field R^{abcd} satisfies $R^{ab(cd)} = R^{a[bcd]} = R^{(ab)cd} = 0$; this in turn implies that $R^{abcd} = R^{cdab}$ (Malament, 2012, p. 258). As usual, the Ricci tensor R_{ab} is defined as $R_{ab} = R^c{}_{abc}$. If, for the derivative operator ∇ associated to a classical spacetime, $R^a{}_{bcd} = 0$ at some $p \in M$, then we say that spacetime is *flat* at p.[10] Similarly, we say that a classical spacetime is *spatially flat* at some $p \in M$ if $R^{abcd} = 0$ at p—for this conditon holds iff parallel transport of spacelike vectors within any spacelike hypersurface $S \subset M$ is, at least locally, path independent: see (Malament, 2012, p. 261). It is worth noting that intermediate between the curvature conditions $R^a{}_{bcd} = 0$ and $R^{abcd} = 0$ is the condition $R^{ab}{}_{cd} = 0$. This holds throughout M iff parallel transport of spacelike vectors along arbitrary curves is, at least locally, path independent (Malament, 2012, p. 263).[11]

To close this subsection, I make some comments on the physical interpretation of classical spacetimes. Again taking the lead from Malament (2012, p. 252),

[10] Since this condition holds iff parallel transport of vectors on M relative to ∇ is, at least locally, path-independent—violations of which constitute the *definition* of spacetime curvature.

[11] Here we still restrict attention to spacelike vectors, but consider their transport along arbitrary curves in M, rather than just curves confined to a particular spacelike hypersurface $S \subset M$.

we can introduce the following interpretive principles.[12] For all smooth curves $\gamma : I \to M$:[13]

- γ is timelike if its image $\gamma[I]$ could be the worldline of a point particle.
- γ can be parameterized so as to be a timelike geodesic (with respect to ∇) iff $\gamma[I]$ could be the worldline of a *free* point particle.
- Clocks record the t_{ab}-length of their worldlines.

If a particle has the image of a timelike curve as its worldline, then we call the tangent field ξ^a of that curve the *velocity* field of the particle, and call $\xi^b \nabla_b \xi^a$ its *acceleration* field. If the particle has a mass m, then its acceleration field satisfies

$$F^a = m\xi^b \nabla_b \xi^a, \tag{4.1.8}$$

where F^a is a spacelike vector field (on the image of its worldline) that represents the net force acting on the particle. This is Newton's second law for NGT and NCT.

4.1.2 Newtonian Gravity

In the four-dimensional, field-theoretic formulation of NGT, models are specified as tuples $\langle M, t_a, h^{ab}, \nabla, \varphi, \Phi \rangle$. The first four elements of this tuple represent classical spacetime structure, as per §4.1.1. In addition, two new elements are introduced in NGT: a real scalar field φ, which represents the *gravitational potential*, and Φ, which is a placeholder for the matter fields of the theory. In NGT, one imposes flatness of ∇ via the field equation[14]

[12] If these principles are taken as fundamental postulates, then the advocate of the dynamical approach to spacetime theories (developed in (Brown and Pooley, 2001, 2006; Brown, 2005)), according to which spacetime structure is a *codification* of certain dynamical properties of matter fields, might demur. Any tension between the following postulates and the dynamical approach is, however, illusory, for the advocate of the dynamical approach can simply regard such points as non-fundamental facts, which can be given some further dynamical explanation.

[13] I is an open interval in $\mathbb{R}$.

[14] On this way of imposing the flatness of ∇ in NGT, the theory is akin to **SR2**—in which flatness of the spacetime metric is imposed dynamically via (3.2.4) (Pooley, 2017, p. 120). As we have seen, there is a theory, **SR1**, corresponding to **SR2**, in which the spacetime metric is fixed to be the Minkowski metric η_{ab} in all KPMs of the theory (i.e. kinematically, rather than dynamically); in **SR1**, unlike **SR2**, the spacetime metric is a *fixed field*. Given existence of such a theory **SR1** corresponding to **SR2**, it should be possible to construct an analogous theory to the above formulation of NGT, in which ∇ is fixed to be flat in this sense. In either case, however, we would expect such fixing of the flatness of ∇ in NGT to mean that there is background structure in the theory; I return to this issue in §4.1.5. Relatedly, one might question why the **SR2**-type formulation of NGT should be taken as the default presentation, rather than the **SR1**-type formulation. The most straightforward answer to this question is that the former presentation renders most explicit the parallels between NGT and NCT, discussed in

$$R^a{}_{bcd} = 0. \tag{4.1.9}$$

A second field equation of NGT is the *Poisson equation*,

$$h^{ab}\nabla_a\nabla_b\varphi = 4\pi\rho. \tag{4.1.10}$$

The mass density function ρ is associated to the Φ in a manner parallel to the relationship between the stress-energy tensor T_{ab} and the matter fields in relativistic spacetime theories.[15] Finally, the gravitational force on a point particle of mass m is given by $-mh^{ab}\nabla_b\varphi$. It follows from (4.1.8) that if the particle is subject to no forces except gravity, and if it has velocity ξ^a, then it satisfies

$$-\nabla^a\varphi = \xi^b\nabla_b\xi^a. \tag{4.1.11}$$

4.1.3 Newton-Cartan Theory

In NCT, the gravitational potential φ in NGT is absorbed into the derivative operator ∇, which is accordingly rendered partly dynamical, and generically curved.[16] Models of NCT are therefore specified by tuples $\langle M, t_a, h^{ab}, \nabla, \Phi\rangle$. The conversion from models of NGT to models of NCT proceeds via Trautman's *geometrization theorem* (Trautman, 1965):[17]

Theorem 1. (Geometrization theorem) *Let $\langle M, t_a, h^{ab}, \nabla\rangle$ be a classical spacetime with ∇ flat ($R^a{}_{bcd} = 0$). Further, let φ and ρ be smooth real-valued functions on M satisfying Poisson's equation $h^{ab}\nabla_a\nabla_b\varphi = 4\pi\rho$. Finally, let $\hat{\nabla} = \left(\nabla, C^a{}_{bc}\right)$, with $C^a{}_{bc} = -t_b t_c \nabla^a\varphi$. Then all of the following hold:*

1. *$\langle M, t_a, h^{ab}, \hat{\nabla}\rangle$ is a classical spacetime.*

§§4.1.3 and 4.1.5. (Finally, recall that converting between such formulations of a theory is not always straightforward—see footnote 39 of Chapter 3.)

[15] Two points are in order here. First, just as it is a mistake in GR to conflate the matter fields Φ and their associated stress-energy tensor T_{ab} (recall footnote 9 in §2.1), so too is it a mistake to conflate the matter fields Φ of NGT and their associated mass density function ρ. This leads to the second point: the mass density function ρ in NGT, as with the stress-energy tensor of GR, affords only a *partial* description of the associated matter fields Φ (cf. §3.6.4). For instance, ρ omits information on the *velocity field* of matter in the theory. (We might then say that while in GR T_{ab} is *partial* and *non-fundamental* (since it may be given a closed-form expression in terms of the matter fields of the theory), in NGT ρ is *partial* and *fundamental*.) As Dewar observes, this particular omission leads to a pernicious (and, until recently, largely unnoticed) indeterminism in NGT—see (Dewar, 2018, p. 4) (here, the reference is to the preprint version of Dewar's paper).

[16] I say *partly*, for the NCT curvature conditions (4.1.14) and (4.1.15) (discussed below), as well as the compatibility conditions (4.1.2) and (4.1.5), imply that not *all* degrees of freedom associated to the derivative operator ∇ of NCT are subject to dynamical evolution. Note also that ∇ being (partly) dynamical does not *entail* that it be generically curved, for it may instead have e.g. torsion or non-metricity degrees of freedom. This latter point will be relevant when I turn to teleparallel gravity below.

[17] This version drawn from (Malament, 2012, pp. 267–268).

2. *$\hat{\nabla}$ is the unique derivative operator on M such that, for all timelike curves on M with four-velocity field ξ^a,*

$$\xi^b \hat{\nabla}_b \xi^a = 0 \iff \xi^b \nabla_b \xi^a = -\nabla^a \varphi \tag{4.1.12}$$

3. *The curvature tensor $\hat{R}^a{}_{bcd}$ associated with $\hat{\nabla}$ satisfies:*

$$\hat{R}_{bc} = 4\pi\rho t_b t_c \tag{4.1.13}$$

$$\hat{R}^{a\ c}_{\ b\ d} = \hat{R}^{c\ a}_{\ d\ b} \tag{4.1.14}$$

$$\hat{R}^{ab}{}_{cd} = 0 \tag{4.1.15}$$

Proof: See (Malament, 2012, pp. 268–269). □

It follows from (4.1.12) that particles subject to a gravitational force in NGT—and thereby exhibiting geodesic deviation with respect to ∇—exhibit geodesic motion with respect to $\hat{\nabla}$. (4.1.13) is the geometrized version of Poission's equation; (4.1.14) holds in a classical spacetime iff this admits, at least locally, a smooth, unit timelike field ξ^a that is geodesic ($\xi^b \nabla_b \xi^a = 0$) and twist-free ($\nabla^{[a}\xi^{b]} = 0$) (Malament, 2012, p. 281); (4.1.15) holds iff parallel transport of spacelike vectors in M is, at least locally, path-independent (Malament, 2012, p. 279).[18] (4.1.13) implies spatial flatness—as Malament writes of this latter result:

> This is striking. It is absolutely fundamental to the idea of geometrised Newtonian theory that *spacetime* is curved (and gravitational is just a manifestation of that curvature). Yet the basic field equation of the theory itself rules out the possibility that *space* is curved. (Malament, 2012, p. 262)

With these points in mind, let me consider now the reverse translation, from NCT to NGT. This proceeds via Trautman's *recovery theorem* (presented by Trautman (1965)):[19]

[18] There is an alternative formulation of NCT due to Künzle (1972, 1976) and Ehlers (1981), in which (4.1.15) is dropped. The motivation for doing so is that this appears to be necessary in order for NCT to qualify as a limiting version of GR (Malament, 2012, pp. 270–271). In this work, I focus on the formulation of NCT due to Trautman (1965), which includes (4.1.15), since in this case a full translation between NGT and NCT is possible. In the Künzle-Ehlers version of NCT, only a translation between NCT and a *generalized* form of NGT is attainable. (In this generalized version of NGT, the gravitational force acting on a particle of unit mass is given by a vector field, but it need not be of the form $\nabla^a\varphi$; in addition, field equations involve a 'rotation field', ω_{ab} (Malament, 2012, pp. 270–271).) Finally, it bears saying here that recently other approaches to the non-relativistic limit of GR have been taken, yielding yet other versions of NCT—I turn to these in more detail below.

[19] This version drawn from (Malament, 2012, pp. 274–275).

Theorem 2. (Recovery theorem) *Let $\langle M, t_a, h^{ab}, \nabla \rangle$ be a classical spacetime satisfying:*

$$R_{ab} = 4\pi\rho t_a t_b \tag{4.1.16}$$

$$R^{a\ c}_{\ b\ d} = R^{c\ a}_{\ d\ b} \tag{4.1.17}$$

$$R^{ab}_{\quad cd} = 0 \tag{4.1.18}$$

for some smooth scalar field ρ on M. Then given any point $p \in M$, there is an open set O containing p, a smooth scalar field φ on O, and a derivative operator $\tilde{\nabla}$ on O such that all the following hold on O:

1. *$\tilde{\nabla}$ is compatible with t_a and h^{ab}.*
2. *$\tilde{\nabla}$ is flat.*
3. *For all timelike curves with four-velocity fields ξ^a,*

$$\xi^b \nabla_b \xi^a = 0 \iff \xi^b \tilde{\nabla}_b \xi^a = -\tilde{\nabla}^a \varphi. \tag{4.1.19}$$

4. *$\tilde{\nabla}$ satisfies Poisson's equation $h^{ab}\tilde{\nabla}_a \tilde{\nabla}_b \varphi = 4\pi\rho$.*

The pair $\langle \tilde{\nabla}, \varphi \rangle$ is not unique. A second pair $\langle \tilde{\nabla}', \varphi' \rangle$, defined on the same open set O, will satisfy the stated conditions iff:

A. $\nabla^a \nabla^b (\varphi' - \varphi) = 0$
B. $\tilde{\nabla}' = \left(\tilde{\nabla}, C'^a_{\ \ bc}\right)$, *where* $C'^a_{\ \ bc} = t_b t_c \nabla^a (\varphi' - \varphi)$.

Proof: See (Malament, 2012, pp. 275–277). □

Thm. 2 tells us that if ∇ arises as the geometrization of the pair $\langle \tilde{\nabla}, \varphi \rangle$, then for any field ψ such that $\nabla^a \nabla^b \psi = 0$, it also arises as the geometrization of the pair $\langle \tilde{\nabla}', \varphi' \rangle$, where $\varphi' = \varphi + \psi$, and $\tilde{\nabla}' = \left(\tilde{\nabla}, t_b t_c \nabla^d \psi\right)$. As Malament (2012, p. 278) points out, this observation captures the fact that, in NGT, the gravitational force on a particle is determined only up to a factor of $m\nabla^a \psi$, where $\nabla^a \psi$ is constant on any one spacelike hypersurface, but can change over time. (This is related to Newton's Corollary VI—see (Saunders, 2013; Knox, 2014; Read and Teh, 2022) for recent discussion.[20]) I return to this observation in §4.1.5.

[20] Recall that Newton's Corollary VI reads as follows: "If bodies are moving in any way whatsoever with respect to one another and are urged by equal accelerative forces along parallel lines, they will all continue to move with respect to one another in the same way as they would if they were not acted on by those forces." (Newton, 2014, p. 99).

4.1.4 Type-II Newton-Cartan Theory

It has been known for some time that it is not straightforward to write down an action principle for NCT (Christian, 1997; Bain, 2004). Hansen et al. (2019a) identify the reasons underlying this difficulty, which have to do with a mathematical inconsistency which arises when one tries to construct such an action, when treating NCT as a gauge theory of non-relativistic gravity (in particular, of the Bargmann group, whose algebra is the central extension of the Poincaré algebra).[21] More significantly, though, Hansen et al. (2019a) also construct a novel version of NCT, which they call 'Type II Newton-Cartan theory' (I denote this henceforth NCT-II), by taking a careful $1/c$ expansion of the Lorentzian metric field g_{ab} of GR; having done this, they *are*, in fact, able to write down an action principle for NCT-II, the Lagrangian for which is (Hansen et al., 2019a, p. 3)

$$\mathscr{L} = -\frac{1}{16\pi G} e\Big[\hat{v}^a \hat{v}^b \check{R}_{ab} - \tilde{\Psi} h^{ab} \check{R}_{ab} - \Psi_{ab} h^{ac} h^{bd} \left(\check{R}_{cd} - a_c a_d - \check{\nabla}_c a_d\right) + \frac{1}{2} \Psi_{ab} h^{ab} \left[h^{cd} \check{R}_{cd} - 2e^{-1} \partial_a \left(e h^{ab} a_b\right)\right]\Big], \quad (4.1.20)$$

where $\hat{v}^a$ is a timelike vector field; $a_a := \mathscr{L}_{\hat{v}} t_a$ is the *torsion vector*; e is the determinant of a Newton-Cartan tetrad field (more on tetrad fields in the next section); $\tilde{\Psi}$ is a scalar field; Ψ_{ab} appears in the $1/c$ expansion of the metric field g_{ab}; and $\check{\nabla}$ is the NCT-II derivative operator.[22] In NCT-II (unlike NCT), the derivative operator $\check{\nabla}$ generically has both torsion and curvature (I explain both of these geometrical properties in the next section).

The details of all the above objects won't matter too much here—readers interested in the details should consult (Hansen et al., 2020), which is a thorough and pedagogical review of this new approach to non-relativistic gravity. For my purposes—which pertain exclusively to the question of the background independence or otherwise of Newton-Cartan theory—the above will suffice.[23]

[21] Going into further detail here would require a substantial detour, so I simply refer the reader to (Hansen et al., 2019a) for the argument.

[22] Hansen et al. (2019a) don't explain what ∂_a is when abstract indices are used, but one can regard this as a 'coordinate derivative operator', in the sense of (Malament, 2012, pp. 61ff.).

[23] There are, however, several further interesting philosophical questions regarding NCT-II. One is that (4.1.20) would seem to afford the possibility for deeper investigations into the notion of gravitational energy in non-relativistic theories—so far, this question has been taken up in e.g. (Dewar and Weatherall, 2018; Dürr and Read, 2019), but progress has been hindered by lack of a variational principle (via which one can define conserved quantities via Noether's theorems, etc.). Second: Hansen et al. (2019b) argue that NCT-II would (perhaps surprisingly) pass many of the classic tests of GR—which, of course, raises the question of what such classic tests in fact *are* testing (this question is taken up by Wolf et al. (2023)).

4.1.5 Background Independence

Newtonian Gravitation

So much for the formal differences between NGT and NCT. What, then, of their background independence? To begin, consider ∇ in NGT. As presented in §4.1.2, flatness of ∇ in NGT is imposed dynamically, via (4.1.9). It would, however, also be possible to impose flatness at the level of KPMs, by restricting such models to those in which ∇ is flat.[24] Though a naïve approach to background independence[25] might deliver the verdict that ∇ counts as background structure in this 'fixed field' formulation of NGT, but not in the formulation presented in §4.1.2, more sophisticated accounts of background independence such as Def. 5 (no absolute objects) and Def. 6 (no formulation with fixed fields) deliver the correct verdict that ∇ counts as background structure in either case.[26]

With this in mind, turn now in earnest to the question of whether NGT is background independent. This theory is sometimes cited as a paragon of background dependence; and, indeed, it is easy to see why: both h^{ab} and t_a are *fixed fields* which, in a Lagrangian formulation of NGT, would not be subject to Hamilton's principle.[27] Due to the presence of such fields, NGT violates Def. 6 (no formulation with fixed fields) and Def. 9 (no absolute fields); due to their non-variational status, NGT also violates Def. 10 (in terms of variational principles) and Def. 15 (in terms of non-trivial corner charges); since such fields qualify as Andersonian absolute objects, NGT violates Def. 5 (no absolute objects).

The existence of such fixed fields leads to there being multiple distinct configurations of the fields $\langle \varphi, \Phi \rangle$ corresponding to the same classical spacetime $\langle M, t_a, h^{ab}, \nabla \rangle$—which I take, quite naturally, to represent *geometrical structure* in the models of NGT.[28, 29] Thus, in Belot's language (§3.6.1), the number of physical degrees of freedom of the theory exceeds the number of geometrical degrees of freedom, so NGT does not satisfy Def. 12.

[24] See footnote 14.

[25] Such as the equation of this notion with the absence of fixed fields—see §3.2.4.

[26] Relatedly, consider the possibility of an alternative formulation of NGT (or, for that matter, NCT), in which the conditions (4.1.1)–(4.1.3) are imposed *dynamically*, via field equations. Doing so might have unattractive consequences, for (unlike e.g. g_{ab} in **SR2**), there exists no straightforward physical interpretation of KPMs where, e.g. the compatibility of the spatial and temporal metrics is violated. This possibility could be explored further in the context of 'Newtonian Weyl geometries', in which the compatibility conditions are relaxed (see (Dewar and Read, 2020)), but the hard work remains to be done to afford such theories a physical interpretation (one expects, for example, 'second clock'-type effects: see (Linnemann and Read, 2021a)).

[27] Classic texts on NGT such as (Friedman, 1983; Malament, 2012) don't identify h^{ab} and t_a explicitly as fixed fields; this, however, appears to me a natural way to proceed, in analogy with **SR1**.

[28] On the form of NGT presented above, while t_a and h^{ab} are fixed in all KPMs, ∇ is equivalent up to isomorphism across DPMs (up to transformations of the form discussed below)—cf. footnote 17 of §3.2.4.

[29] Below, I modify the assumption that $\langle M, t_a, h^{ab}, \nabla \rangle$ constitutes geometrical structure in a model of NGT.

While it is straightforward to issue the above verdict on NGT, the situation regarding Belot's account is, in fact, rather delicate. First, one must consider whether NGT satisfies Belot's Def. 11, of *full* background dependence. Recall from Prop. 2 that the derivative operator ∇ of NGT is not unique; thus, there are models of the theory with the same $\langle \varphi, \Phi \rangle$, but set in classical spacetimes differing in their derivative operators. In more detail: for a given DPM of NGT $\langle M, t_{ab}, h^{ab}, \nabla, \varphi, \Phi \rangle$, $\langle M, t_{ab}, h^{ab}, \nabla', \varphi, \Phi \rangle$ is also a DPM of NGT, where $\nabla' = \left(\nabla, C^{b}{}_{cd}\right)$, $C^{a}{}_{bc} = 2h^{ad} t_{(b}\kappa_{c)d}$, and $\nabla_{[a}\kappa_{b]c} = 0$.[30, 31] (For further discussion of κ_{ab}, see (Bekaert and Morand, 2016; Read and Teh, 2018; Teh, 2018).) A schematic response to this result is the following: 'The geometries of such models (are/are not) equivalent, and the models [do/do not] exhibit gauge redundancy'. Since, however, it is not plausible to understand such models as having equivalent geometries yet as not being gauge equivalent (since these models are identical in all other respects), I turn now to the other three possibilities.

First, suppose that the geometries of such models *are* equivalent, and the models *are* considered gauge equivalent. In that case, in this regard, to each physical configuration, there corresponds a unique geometrical configuration—so such a redundancy does not inhibit NGT's being background independent on Belot's account.[32] Second, suppose that the geometries of such models are *not* equivalent, and the models are *not* considered gauge equivalent. Then, it is still the case that (in this regard) each distinct model corresponds to a unique geometry, so again, the background independence of NGT on Belot's account is not threatened.[33] Third, consider the possibility that such geometries are *not* equivalent, but the models in question *are* gauge equivalent.[34] In that case, Def. 12 is straightforwardly not satisfied. Thus, on this latter view, there exists an additional sense in which NGT violates Def. 12.

On the second and third interpretive routes above, that the objects of the theory (and in particular ∇) might be inequivalent across DPMs means that it is *not* the case that there are *no* geometrical degrees of freedom in the theory. Therefore, it appears on such interpretations that one *cannot* regard NGT as *fully* background dependent, on Def. 11.

[30] One uses (4.1.6) to show that (4.1.10) is invariant under this change of derivative operator; to show that (4.1.9) is also invariant, it is convenient to use $R'^{a}{}_{bcd} = R^{a}{}_{bcd} + 2\nabla_{[c}C^{a}{}_{d]b} + 2C^{e}{}_{b[c}C^{a}{}_{d]e}$ (Malament, 2012, p. 70), where $R'^{a}{}_{bcd}$ is the Riemann tensor associated to ∇'.

[31] Imposing flatness of ∇ at the level of KPMs (as discussed above) does not dissolve this multiplicity, unless one restricts at the level of KPMs to *one* such choice of ∇.

[32] Though recall that the background independence of NGT according to Def. 12 fails for other reasons—namely, that h^{ab} and t_a are fixed across all models.

[33] Though see footnote 32.

[34] Such a view may be defensible on the interpretational approach to symmetries, but, absent some account of *how* such models with inequivalent geometries can be gauge equivalent, is not immediately defensible on the motivational approach (cf. §2.2). Note also that if one takes 'equivalent geometries' to mean *physically* equivalent geometries, then it is questionable whether two gauge-equivalent models can possess inequivalent geometries—in which case, this third option is excluded.

Indeed, a similar point arises out of a second (more familiar) multiplicity in the models of NGT. To see this, recall from Thm. 2 that for a given model with derivative operator and gravitational potential $\langle \nabla, \varphi \rangle$,[35] there exists a class of other models of NGT with $\langle \nabla', \varphi' \rangle$—where $\varphi' = \varphi + \psi$; $\nabla' = \left(\nabla, t_b t_c \nabla^d \psi\right)$; and ψ is a field such that $\nabla^a \nabla^b \psi = 0$—which correspond to the same model of NCT. This being so, it is argued in e.g. (Malament, 2012; Knox, 2014; Weatherall, 2016a) that it is plausible to understand these as being gauge equivalent.[36]

In this case, we face the same four interpretive options as above. First, to regard such classes of models as having equivalent geometries yet not being gauge equivalent is now a live option, since the φ field now differs from model to model within such classes. As above, one may also argue that such models have inequivalent geometries, but *are* gauge equivalent. In either case, the biconditional of Def. 12 is violated, so such an interpretation again yields a further sense in which NGT is not background independent, on Belot's definition. Alternatively, if one regards all such models as having equivalent geometries and as being gauge equivalent (*à la* (Malament, 2012, p. 278) and (Weatherall, 2016a, §6)), or as having inequivalent geometries and as not being gauge equivalent, then this multiplicity of models does not give rise to further roadblocks to NGT being background independent, on Def. 12. Again, however, any interpretation here on which the derivative operators are *not* equivalent across models results in it *not* being the case that there are *no* geometrical degrees of freedom in NGT—thereby posing problems for any attempt to understand this theory as fully background dependent, as per Def. 11.

The intricacies regarding the interplay between NGT and Belot's account do not end here. Consider now Knox's suggestion that "the spacetime structure of NGT is represented not by the flat . . . connection usually made explicit in formulations, but by the sum of the flat connection and the gravitational field" (Knox, 2014, p. 863) (cf. §3.8). If one takes this seriously, then one must understand the geometrical structure in models of NGT to be picked out not by $\langle M, t_{ab}, h^{ab}, \nabla \rangle$, but by $\langle M, t_{ab}, h^{ab}, \nabla, \varphi \rangle$. Given such a tuple, a unique ρ is picked out via (4.1.10); moreover, given a different tuple $\langle M, t_{ab}, h^{ab}, \nabla', \varphi' \rangle$, the *same* ρ is picked out, via the above reasoning. If one understands such models to have equivalent

[35] Having selected such a derivative operator, as per the above discussion.

[36] On this multiplicity of models in NGT, Malament (2012, p. 279) points out that if we are dealing with a bounded mass distribution—i.e. if ρ has compact support on every spacelike hypersurface—and moreover if we require that the gravitational field fall off as one approaches spatial infinity, then the multiplicity disappears. The reason for this is that if $\nabla^a \psi$ is constant on spacelike hypersurfaces *and* goes to zero at spatial infinity, then it must vanish everywhere. Thus, by imposing such boundary conditions, this multiplicity collapses, and cannot further threaten the background independence of NGT according to Def. 12. (Cf. footnote 52 in Chapter 3, in which I mentioned a related point in the context of GR.)

geometries,[37] and to be gauge equivalent,[38] then it is clear that NGT *does* satisfy Def. 12.[39]

This result is important, for intuitively it is the incorrect verdict—the reason being that in *every* model of NGT, h^{ab} and t_b *remain fixed*. The problem stems from the fact that Belot's account of background independence does not adequately account for the possibility of *multiple* geometric objects in a theory being counted as geometrical structure in the sense of §3.6.1.[40] To overcome this problem, a caveat must be added to Belot's account: if multiple geometric objects in the theory are considered to be geometrical structure, they must *all* be inequivalent in all non-gauge equivalent models of the theory. Thus, we see that NGT presents a novel and concrete problem case for Belot's account, which can be used to tune the program.

My ultimate verdict on the background independence of NGT is, then, as follows: while the theory violates Def. 5 (no absolute objects), Def. 6 (no formulation in terms of fixed fields), Def. 9 (no absolute fields), Def. 10 (variational principles), and Def. 15 (non-trivial corner charges), a number of subtleties inhibit a clear-cut conclusion on Def. 11 (Belot's proposal). These results in hand, I close by turning to whether NGT is diffeomorphism invariant on Def. 3. The straightforward answer here is *no*, for, treating h^{ab} as a fixed field and performing an arbitrary smooth coordinate transformation (implemented via a diffeomorphism d) on (4.1.10), one extracts the condition for the associated transformation $d^* h^{ab} = h^{ab}$. When the transformation is an affine transformation, this restricts to the Galilean transformations; when one drops the condition that the transformation be affine, the restriction is to the Leibniz transformations.[41] Transparently, then, this theory is *not* diffeomorphism invariant.

Newton-Cartan Theory

This discussion of NGT in hand, I turn now to NCT. Since, in this theory, gravitation is 'geometrized' in a manner similar to GR,[42] one might expect NCT also to be background independent. This, however, is *not* the case, for both h^{ab} and t_a *remain fixed*. While ∇ is rendered *partly* dynamical,[43] such fixing of the geometric objects of NCT renders the theory background dependent, on Def. 5

[37] Which is plausible, if, with Knox, one considers only the *sum* of contributions from a given $\langle \nabla', \varphi' \rangle$ to be physical (Knox, 2011, p. 870).

[38] Which, as mentioned above, is a mainstream position.

[39] Throughout this section, I set aside further complications regarding the 'pernicious indeterminism' of footnote 15.

[40] Below, I find the same result in the case of NCT. This is not surprising, for classes of models of NGT differing in $\langle \nabla, \varphi \rangle$ correspond to the *same* model of NCT (cf. §4.1.3).

[41] For the definitions of these transformations, see (Pooley, 2013).

[42] Although recall footnote 5.

[43] But still not *fully*, in light of curvature conditions (4.1.14) and (4.1.15) and metric compatibility (4.1.2) and (4.1.5)—recall footnote 16. See also (Norton, 1993, p. 828) for relevant comments.

(no absolute objects), Def. 6 (no formulation in terms of fixed fields), Def. 9 (no absolute fields), Def. 10 (variational principles),[44] and Def. 15 (non-trivial corner charges).

It might, in addition, be tempting to state that such fixing of the geometric objects of NCT renders the theory background dependent on Belot's account. This, however, is not correct. Recall that for full background independence, Belot requires a one-one match between the geometrical degrees of freedom of the theory—which here parameterize the difference between tuples $\langle t_a, h^{ab}, \nabla \rangle$ across models—and the physical degrees of freedom—which here parameterize the difference between tuples $\langle t_a, h^{ab}, \nabla, \Phi \rangle$ across gauge inequivalent models. However,[45] since the Ricci tensor associated to the derivative operator ∇ and the mass density function ρ of the matter fields Φ are coupled via the field equation (4.1.13), on Belot's account it turns out that there *does* exist a unique geometry corresponding to each distinct physical configuration—meaning that NCT *does* satisfy Def. 12.

As in the case of NGT, the above is an intuitively incorrect result. However, by introducing the above-mentioned caveat that, if multiple geometric objects in the theory are considered to be geometrical structure, they must *all* be inequivalent in all non-gauge equivalent models of the theory, one may circumvent this verdict.

Let me close this assessment of NCT with three comments—the first two are easy to make; the third requires a more involved discussion. First, the non-uniqueness of the derivative operator ∇ in a classical spacetime (c.f. §4.1.1) may, depending on one's interpretation, threaten the background independence of NCT in exactly the same manner as in the case of NGT, discussed above. Second, like NGT, NCT is not diffeomorphism invariant according to Def. 3—to see this, consider the fact that the left-hand side of (4.1.13) is composed of dynamical objects, and so will transform under diffeomorphisms; on the other hand, since the t_a on the right-hand side of (4.1.13) are fixed fields, one does not (recalling again Def. 3) transform these fields in assessments of diffeomorphism invariance; putting this together, however, there is no guarantee that the appropriately transformed fields still satisfy (4.1.13).

The third comment is this. Insofar as Newtonian theories involve a natural decomposition between space and time (given the specification in the KPMs of these theories of h^{ab} and t_a and the orthogonality relation (4.1.1) obtaining between these objects), the possibility is opened for distinguishing *spatial* background independence from *temporal* background independence from background

[44] In fact, given that we have already seen above that NCT lacks a Lagrangian formulation, the question of whether Def. 10 is satisfied is blocked before one even need consider whether or not the fields are subject to Hamilton's principle.

[45] Modulo concerns analogous to those regarding GR, as presented in §3.6.4. Cf. also footnote 39.

independence *tout court*. By this I mean the following: recall that the Newtonian theories considered above are spatially flat, meaning that the Riemann tensor of the relevant derivative operator (whether that of NGT or that of NCT), when projected onto spatial hypersurfaces, is flat (Malament, 2012, ch. 4). Given this, let us say that a geometric object in a Newtonian theory qualifies as spatial 'background structure' when its projection onto spatial hypersurfaces violates one or more of our definitions of background independence. For example, the projection of a derivative operator associated with a classical spacetime in a Newtonian theory does not vary from DPM to DPM (it is always flat); therefore, this would qualify as spatial 'background structure', given Def. 5. Likewise, a geometric object in a Newtonian theory qualifies as temporal 'background structure' when its projection onto the temporal dimension (orthogonal to the spatial hypersurfaces) violates one or more of our definitions of background independence. Although an in-principle useful distinction (also applicable to other theories, for example those discussed in §4.4), in fact, the same verdicts as above (summarized in Table 4.1, presented at the end of this chapter) hold true both in the case of spatial and of temporal background independence in the Newtonian theories considered in this chapter. The only additional interesting comment to be made is the following: when it comes to spatial background independence on Belot's account—Def. 12—the account does not face the issue of giving the intuitively incorrect verdict on background independence in cases in which some pieces of geometrical structure vary from model to model but others do not: for the spatial projections of *all* elements of a classical spacetime geometry do not change (up to isomorphism) from DPM to DPM of the Newtonian theories considered here, given spatial flatness. On the other hand, Belot's account retains this problem in the case of temporal background independence in NCT, as the derivative operator of that theory can have non-trivial temporal curvature, which can be inequivalent from model to model.

Type-II Newton-Cartan Theory

Finally, let me say something on the background independence of NCT-II—though these remarks will be somewhat tentative, given the nascent status of that theory. Since h^{ab} and t_a are obtained in this case by taking a non-relativistic limit of g_{ab} (namely, a $1/c$ expansion of that metric), it is most natural to view these objects as being dynamical and variational in this case (unlike in NGT or NCT)—indeed, they are treated as such by Hansen et al. (2019a, p. 4).

When the dynamical status of h^{ab} and t_a is assumed, one finds that verdicts on background independence for NCT-II are remarkably different from those given for NGT or NCT. On Def. 6 (no formulation in terms of fixed fields) and Def. 9 (no absolute fields), the theory turns out to be background independent (modulo, of course, the possibility that one might be able to reformulate NCT-II in terms of fixed fields, which would lead to a negative verdict in the former case—the proof here, however, is in the pudding). On Def. 5 (no absolute objects), the situation becomes a little more involved: does the square root of the determinant of h^{ab}, say, count as an absolute object in this case?[46] In fact, perhaps surprisingly, the answer is interesting: since h^{ab} is *degenerate*, such an object is identically zero! One might think that this affords good ground for stating that Def. 5 is *not* violated; on the other hand, since this determinant is still *well-defined*, one might maintain the line that if (the analogue of) this object leads to violations of Def. 5 in the context of GR, then it should also lead to violations of Def. 5 in the context of NCT-II. This I regard as being the most consistent application of this definition.

Turning now to Def. 10, it at least appears that all of the objects in (4.1.20) have physical significance ($\hat{v}^a$, for example, represents a congruence of physical observers; regarding the physical significance of $\tilde{\Psi}$ and Ψ_{ab}, since these objects are obtained from the $1/c$ expansion of g_{ab}—see (Hansen et al., 2019a, p. 3)—one expects them to have *some* physical significance[47]), in which case Def. 10 (variational principles) is also satisfied. More work would need to be done in order to demonstrate the background independence of NCT-II on Def. 15 (non-trivial corner charges), but here too there is reason to be optimistic about an affirmative answer. Just as in the case of **SR2** or **GR1**, one expects the action (which is generally covariant, thereby satisfying the first clause of Def. 15) to encode global degrees of freedom, manifesting as non-trivial corner charges.[48]

The prospects for satisfaction of Belot's Def. 12 also appear improved in NCT-II, since (as explained above) the most straightforward understanding of this version of the theory involves all objects being dynamical, and given the prospects for the introduction of couplings between geometrical and dynamical degrees of freedom in the associated Lagrangian. Thus, NCT-II appears to be the version of non-relativistic gravity most akin to GR in terms of its satisfaction of Belot's definition of background independence: while there may be further roadblocks to this verdicts depending upon the nature of the material fields (as, for example, encountered for Maxwell theory in GR in §3.6.4), abstracting away from the details of particular cases suggests the tenability of a verdict of full background independence here.

[46] Cf. §3.3.3.

[47] Indeed, $\tilde{\Psi}$ can be understood as a version of the Newtonian potential. Hansen et al. (2019a) are not explicit about the physical significance of Ψ_{ab}, and arguably there remains more to be said regarding what exactly this object is supposed to represent.

[48] Of course, in this case one should actually perform the calculation to confirm this. Given the complexity of (4.1.20), this will have to remain a task for another day.

4.2 General Relativity and Teleparallel Gravity

NCT is the geometrized theory of gravity associated with the force theory NGT. The theoretical equivalence[49] between NGT and NCT raises the question: is there a force theory associated to GR? As Knox (2011) states, it appears that the answer here is *yes*, the theory in question being *teleparallel gravity* (TPG). In TPG, the Levi-Civita connection of GR, which has curvature but no torsion, is replaced with a *Weitzenböck connection*, with torsion but no curvature.[50] In this section, I review the details of GR (§4.2.1) and TPG (§4.2.2). I then present explicit translation theorems relating the two theories, in the style of §4.1.3. This done, I assess in §4.2.3 the background independence of TPG. Given the analogy between NGT and NCT, and TPG and GR, it is natural to ask: is the pattern of judgments regarding the background independence of NGT and NCT preserved in the case of TPG and GR?

4.2.1 General Relativity

Let us recall the essential details of GR.[51] First, define a *relativistic spacetime* to be a pair $\langle M, g_{ab} \rangle$, where M is a four-dimensional differentiable manifold, and g_{ab} is a Lorentzian metric field on M. Associated to $\langle M, g_{ab} \rangle$, there is a unique derivative operator ∇, which is: (i) torsion free—in the sense that the associated *torsion tensor* $T^a{}_{bc}$, defined through $T^a{}_{bc} X^b Y^c := \nabla_b X^b Y^a - X^a \nabla_b Y^b - [X, Y]^a$, vanishes—and (ii) metric compatible—in the sense that $\nabla_a g_{bc} = 0$. Given the uniqueness of this derivative operator, ∇ is not included in models of GR, which are denoted by triples $\langle M, g_{ab}, \Phi \rangle$, where Φ is again a placeholder for the matter fields of the theory.[52, 53] The dynamical equations of GR are the Einstein equation, (2.1.1), plus dynamical equations associated to the matter fields.[54]

[49] See footnote 4.

[50] In contemporary versions of TPG, a 'generalized Weitzenböck connection', defined via the fibre bundle formalism in terms of the spin connection, is used. The details will not matter for my discussion in this section, and hence I focus on the original version of TPG. For further details on this modern, 'covariant' version of TPG, see (Aldrovandi and Pereira, 2013; Read and Teh, 2018).

[51] See also §2.1.

[52] Clearly, the existence of such a unique derivative operator is in contrast with NGT and NCT, for which we saw in §§4.1.2 and 4.1.3 that no *unique* such operator exists.

[53] If one were to so desire, one could introduce ∇ as separate element of one's KPMs, and then have compatibility enforced dynamically—this is the essence of Palatini's approach, surveyed in e.g. (Wald, 1984, pp. 454–455). One can also generalize the Palatini approach so that torsion-freeness is derived dynamically rather than being assumed *ab initio*; the result is what is known as *Einstein-Cartan theory* (see (Trautman, 2006) for an introduction).

[54] See also footnote 10 of §2.1.

4.2.2 Teleparallel Gravity

Tetrad Fields

Knox (2011) claims that TPG is to GR what NGT is to NCT.[55] To present TPG, I must first define a *tetrad field*. Suppose that at some $p \in M$, we are working within the constraints of a coordinate basis for the tangent space T_pM at p. Generically, such coordinates will not be orthogonal. Nevertheless, at each $p \in M$ we can always find an orthonormal basis for T_pM: for every $p \in M$, T_pM has a set of four basis vectors $h_{\mathsf{a}}{}^{a}$ ($\mathsf{a} = 1 \dots 4$) such that

$$g_{ab}h_{\mathsf{a}}{}^{a}h_{\mathsf{a}}{}^{b} = \begin{cases} +1 & \text{if } \mathsf{a} = 1 \\ -1 & \text{if } \mathsf{a} = 2,3,4 \end{cases} \tag{4.2.1}$$

and $g_{ab}h_{\mathsf{a}}{}^{a}h_{\mathsf{b}}{}^{b} = 0$ if $\mathsf{a} \neq \mathsf{b}$. In other words, at each $p \in M$, we can always find a set of four basis vectors $h_{\mathsf{a}}{}^{a}$ for T_pM such that

$$g_{ab}h^{a}{}_{\mathsf{a}}h^{b}{}_{\mathsf{b}} = \eta_{\mathsf{ab}}. \tag{4.2.2}$$

With this in mind, define a *tetrad field* $h^{a}{}_{\mathsf{a}}$ to be a set of four vector fields which at each $p \in M$ provide such an orthonormal basis for T_pM. In (4.2.2) and what follows, $(\mathsf{a}, \mathsf{b}, \dots)$ are understood to be indices in an orthonormal basis provided by the tetrad field.[56] Importantly, it follows as a lemma from (4.2.2) that[57]

$$\eta_{\mathsf{ab}}h^{\mathsf{a}}{}_{a}h^{\mathsf{b}}{}_{b} = g_{ab}, \tag{4.2.3}$$

$$h^{\mathsf{a}}{}_{a}h^{b}{}_{\mathsf{a}} = \delta^{b}{}_{a}. \tag{4.2.4}$$

The Weitzenböck Derivative Operator and Teleparallel Gravity

Given a tetrad field, we define the components $\bar{\Gamma}^{\rho}{}_{\mu\nu}$ of the *Weitzenböck derivative operator* $\bar{\nabla}$ to be (de Andrade et al., 2000, p. 3)

$$\bar{\Gamma}^{\rho}{}_{\mu\nu} := h^{\rho}{}_{\mathsf{a}}\partial_{\nu}h^{\mathsf{a}}{}_{\mu}. \tag{4.2.5}$$

[55] Although 'teleparallel gravity' can be used to refer to a family of theories using the Weitzenböck connection, the variant discussed here reproduces the results of GR exactly. Note also that the analogy between the GR-TPG relationship on the one hand, and the NCT-NGT relationship on the other, has been made mathematically precise in (Read and Teh, 2018; Schwartz, 2022); see also (Read and Teh, 2022).

[56] More technically, while Roman indices are tangent space T_pM indices, sans serif indices are valued in a Minkowski vector space associated with each $p \in M$. For further details, see (Wallace, 2015; Weatherall, 2016b).

[57] To prove (4.2.3), consider $\eta_{\mathsf{ab}}h^{\mathsf{a}}{}_{a}h^{\mathsf{b}}{}_{b}h^{b}{}_{\mathsf{c}} = \eta_{\mathsf{ab}}h^{\mathsf{a}}{}_{a}\delta^{\mathsf{b}}{}_{\mathsf{c}} = \eta_{\mathsf{ac}}h^{\mathsf{a}}{}_{a} = h_{\mathsf{c}a} = g_{ab}h^{b}{}_{\mathsf{c}}$; cancelling the $h^{b}{}_{\mathsf{c}}$ yields the result. To prove (4.2.4), simply raise the b index in (4.2.3).

$\bar{\nabla}$ is the derivative operator of TPG.[58] This in hand, I state that KPMs of TPG are triples $\langle M, h^{\mathsf{a}}{}_{a}, \Phi \rangle$, where $h^{\mathsf{a}}{}_{a}$ is a tetrad field on M, and, as before, Φ is a placeholder for the matter fields of the theory. Analogously with GR (and in contrast with NGT and NCT), the Weitzenböck derivative $\bar{\nabla}$ is not included as a separate element in the model, for, given a tetrad field $h^{\mathsf{a}}{}_{a}$, there is a unique such operator, defined via (4.2.5). Note also that, on this understanding of TPG, a metric field is not part of the basic kinematical structure of the theory.

Non-Uniqueness of the Tetrad Fields

For any $p \in M$, the set of tetrad fields $h^{\mathsf{a}}{}_{a}$ for T_pM is not unique. In fact, (4.2.2) only determines the tetrad field up to local Lorentz transformations (Aldrovandi and Pereira, 2013, p. 13). To see this, suppose that there exists another tetrad field $h'^{\mathsf{a}}{}_{a}$, also satisfying (4.2.2). Contracting both sides of (4.2.3) with $h'_{\mathsf{c}}{}^{a} h'_{\mathsf{d}}{}^{b}$ and using (4.2.2) in terms of $h'^{\mathsf{a}}{}_{a}$, we obtain

$$\eta_{\mathsf{ab}} h'_{\mathsf{c}}{}^{a} h_{a}{}^{\mathsf{a}} h'_{\mathsf{d}}{}^{b} h_{b}{}^{\mathsf{b}} = \eta_{\mathsf{cd}}. \tag{4.2.6}$$

Defining $M_{\mathsf{a}}{}^{\mathsf{b}} := h'_{\mathsf{a}}{}^{a} h_{a}{}^{\mathsf{b}}$, we have

$$\eta_{\mathsf{ab}} M_{\mathsf{c}}{}^{\mathsf{a}} M_{\mathsf{d}}{}^{\mathsf{b}} = \eta_{\mathsf{cd}}, \tag{4.2.7}$$

so the $M_{\mathsf{a}}{}^{\mathsf{b}}$ satisfy the definition of Lorentz matrices. Rearranging the definition of $M_{\mathsf{a}}{}^{\mathsf{b}}$, we have

$$h'^{\mathsf{a}}{}_{a} = M^{\mathsf{a}}{}_{\mathsf{b}} h^{\mathsf{b}}{}_{a}. \tag{4.2.8}$$

Thus, any field $h'^{\mathsf{a}}{}_{a}$ which differs from a tetrad field $h^{\mathsf{a}}{}_{a}$ by a Lorentz transformation will also be a tetrad field. Considering geometric objects of TPG such as $\bar{\Gamma}^{\rho}{}_{\mu\nu}$ and $\bar{T}^{\rho}{}_{\mu\nu}$,[59] it is then straightforward to verify that such objects are *invariant* under such Lorentz transformations of tetrad indices (Aldrovandi and Pereira, 2013, p. 13).[60]

The Levi-Civita Derivative

In presentations of TPG, reference is sometimes made to the *Levi-Civita derivative operator*. The referent of this term should, however, be made explicit, for unlike

[58] Or at least, is the derivative operator of TPG as traditionally understood—see footnote 50.

[59] $\bar{T}^{a}{}_{bc}$ is the torsion tensor (defined in §4.2.1) associated to $\bar{\nabla}$; this object should not be conflated with the stress-energy tensor of TPG, discussed below.

[60] $\bar{\Gamma}'^{\rho}{}_{\mu\nu} = h'^{\rho}{}_{\mathsf{a}} \partial_{\nu} h'^{\mathsf{a}}{}_{\mu} = (M^T M)_{\mathsf{a}}{}^{\mathsf{b}} h^{\rho}{}_{\mathsf{b}} \partial_{\nu} h^{\mathsf{a}}{}_{\mu} = \delta_{\mathsf{a}}{}^{\mathsf{b}} h^{\rho}{}_{\mathsf{b}} \partial_{\nu} h^{\mathsf{a}}{}_{\mu} = h^{\rho}{}_{\mathsf{a}} \partial_{\nu} h^{\mathsf{a}}{}_{\mu} = \Gamma^{\rho}{}_{\mu\nu}$. Since $\bar{T}^{\rho}{}_{\mu\nu}$ is defined directly from $\bar{\Gamma}^{\rho}{}_{\mu\nu}$, this object, and in turn the force and field equations of TPG ((4.2.21) and (4.2.29)—discussed in §4.2.2), are invariant under such transformations. Indeed, as stated at (Aldrovandi and Pereira, 2013, p. 13), *all* spacetime-indexed quantities in TPG should be so invariant.

GR, no Lorentzian metric field is fundamental in TPG. To do so, first recall that, generically, the coefficients $\Gamma^{\rho}{}_{\mu\nu}$ of the Levi-Civita derivative operator associated with some Lorentzian metric field g_{ab} are given by

$$2g_{\sigma\rho}\Gamma^{\rho}{}_{\mu\nu} = \partial_{\mu}g_{\sigma\mu} + \partial_{\mu}g_{\sigma\nu} - \partial_{\sigma}g_{\mu\nu}. \tag{4.2.9}$$

Using (4.2.3), the right hand side of (4.2.9) can be expressed in terms of the tetrad field as

$$2g_{\sigma\rho}\Gamma^{\rho}{}_{\mu\nu} = \eta_{\mathrm{ab}}h^{\mathrm{b}}{}_{\mu}\partial_{\nu}h^{\mathrm{a}}{}_{\sigma} + \eta_{\mathrm{ab}}h^{\mathrm{a}}{}_{\sigma}\partial_{\nu}h^{\mathrm{b}}{}_{\mu} + \eta_{\mathrm{ab}}h^{\mathrm{b}}{}_{\nu}\partial_{\mu}h^{\mathrm{a}}{}_{\sigma} + \\ \eta_{\mathrm{ab}}h^{\mathrm{a}}{}_{\sigma}\partial_{\mu}h^{\mathrm{b}}{}_{\nu} - \eta_{\mathrm{ab}}h^{\mathrm{b}}{}_{\nu}\partial_{\sigma}h^{\mathrm{a}}{}_{\mu} - \eta_{\mathrm{ab}}h^{\mathrm{a}}{}_{\mu}\partial_{\sigma}h^{\mathrm{b}}{}_{\nu}. \tag{4.2.10}$$

(4.2.10) can be considered a definition of the components of the Levi-Civita derivative in TPG in terms of the tetrad field.[61] Given this, it is useful to compute the difference of the components of the Weitzenböck and Levi-Civita derivatives; we find

$$\bar{\Gamma}^{\rho}{}_{\mu\nu} - \Gamma^{\rho}{}_{\mu\nu} = K^{\rho}{}_{\mu\nu}, \tag{4.2.11}$$

where $K^{\rho}{}_{\mu\nu}$ is the *contorsion tensor* (de Andrade et al., 2000, p. 3), defined as

$$K^{a}{}_{bc} = \frac{1}{2}\left(\bar{T}_{b}{}^{a}{}_{c} + \bar{T}_{c}{}^{a}{}_{b} - \bar{T}^{a}{}_{bc}\right). \tag{4.2.12}$$

The Teleparallel Force Law

TPG is supposed to be "a gauge theory for the translation group" (de Andrade et al., 2000, p. 1). In order to implement this, first consider a frame in which $h^{\mathrm{a}}{}_{\mu} = \partial_{\mu}x^{\mathrm{a}}$;[62] then augment the tetrad field with an extra term $B^{\mathrm{a}}{}_{\mu}$: (Pereira, 2014, §3.1)

$$h^{\mathrm{a}}{}_{\mu} = \partial_{\mu}x^{\mathrm{a}} + B^{\mathrm{a}}{}_{\mu}. \tag{4.2.13}$$

Inspired by this decomposition, one then considers a translation of the 'internal' (Minkowski vector space) coordinates, $x^{\mathrm{a}} \to x^{\mathrm{a}} + \Delta^{\mathrm{a}}$, under which, generically, we can write $B^{\mathrm{a}}{}_{\mu} \to B'^{\mathrm{a}}{}_{\mu}$. After such a transformation, the tetrad field becomes

$$h^{\mathrm{a}}{}_{\mu} \to h'^{\mathrm{a}}{}_{\mu} = \partial_{\mu}x^{\mathrm{a}} + \partial_{\mu}\Delta^{\mathrm{a}} + B'^{\mathrm{a}}{}_{\mu}. \tag{4.2.14}$$

[61] Strictly, for this to be a definition one should also rewrite the $g_{\sigma\rho}$ on the left hand side of (4.2.10) in terms of the tetrad field, via (4.2.3).

[62] This is possible locally in light of (3.5.4). Indeed, the analogue between (3.5.4) and (4.2.3) highlights the close connection between tetrad fields and clock fields.

Under what circumstances does $h'^{a}{}_{\mu}$, as with $h^{a}{}_{\mu}$, satisfy the definition of the tetrad field (4.2.2)? This result holds just if

$$B^{a}{}_{\mu} \to B'^{a}{}_{\mu} = B^{a}{}_{\mu} - \partial_{\mu}\Delta^{a}. \tag{4.2.15}$$

In the formulation of TPG under consideration here, we *stipulate* that $B^{a}{}_{\mu}$ has such transformation properties under a change $x^{a} \to x^{a} + \Delta^{a}$; we then see that the tetrad field $h^{a}{}_{\mu}$ is an *invariant object* under such transformations (Aldrovandi and Pereira, 2013, p. 38).[63, 64] Inspired by (4.2.15), it is natural to view TPG as an Abelian gauge theory of the translation group. With this in mind, we define the components of an Abelian field strength tensor (de Andrade et al., 2000, p. 2)

$$\begin{aligned} F^{a}{}_{\mu\nu} &:= \partial_{\mu}B^{a}{}_{\nu} - \partial_{\nu}B^{a}{}_{\mu} & (4.2.16)\\ &= \partial_{\mu}h^{a}{}_{\nu} - \partial_{\nu}h^{a}{}_{\mu}; & (4.2.17) \end{aligned}$$

the second line here follows using (4.2.14). Now, in a coordinate basis, the definition of the torsion tensor (see §4.2.1) reads $T^{\lambda}{}_{\mu\nu} = \Gamma^{\lambda}{}_{\mu\nu} - \Gamma^{\lambda}{}_{\nu\mu}$. In the case of the Weitzenböck connection, we hence have

$$\begin{aligned} \bar{T}^{\lambda}{}_{\mu\nu} &= \bar{\Gamma}^{\lambda}{}_{\mu\nu} - \bar{\Gamma}^{\lambda}{}_{\nu\mu} \\ &= h_{a}{}^{\lambda}\partial_{\nu}h^{a}{}_{\mu} - h_{a}{}^{\lambda}\partial_{\mu}h^{a}{}_{\nu}. & (4.2.18) \end{aligned}$$

Multiplying through by $h^{b}{}_{\lambda}$, it follows that

$$h^{a}{}_{\lambda}\bar{T}^{\lambda}{}_{\mu\nu} = \partial_{\mu}h^{a}{}_{\nu} - \partial_{\nu}h^{a}{}_{\mu}, \tag{4.2.19}$$

wherefrom we obtain the result

$$F^{a}{}_{\mu\nu} = h^{a}{}_{\lambda}\bar{T}^{\lambda}{}_{\mu\nu}. \tag{4.2.20}$$

[63] In this language, $h^{a}{}_{\mu}$ is *gauge-invariant* (Aldrovandi and Pereira, 2013, p. 38), *pace* Knox (2011). As a consequence, Knox's further claim that "the tetrad field is defined only up to a local gauge transformation, and this gauge freedom passes onto the connection defined in terms of the tetrads" (Knox, 2011, p. 273) is also incorrect. While it is possible that Knox has in mind here the non-uniqueness of the tetrad field discussed in §4.2.2, there also exist problems on this reading. First, while it may also be plausible to treat such Lorentz transformations of the tetrad fields as gauge transformations, it is confusing to run together these two possible senses of 'gauge transformation' in TPG without qualification. Second, even putting this aside, the claim that "this gauge freedom passes onto the connection defined in terms of the tetrads" is incorrect, for we have seen that the components of the Weitzenböck derivative are formally *invariant* under such a transformation.

[64] By augmenting $h^{a}{}_{\mu}$ with $B^{a}{}_{\mu}$, we are essentially 'dressing' the field so as to be invariant under local gauge transformations (here, translations). For more on the dressing field approach to symmetries, see (François, 2019; Wolf et al., 2022).

As in §4.1, defining a particle four-velocity to be ξ^a, and recalling that the Lorentz force law[65] in covariant notation reads $\xi^a \nabla_a \xi^b = F^b{}_a \xi^a$,[66] we define by analogy the force law of TPG to be

$$\begin{aligned} \xi^a \bar{\nabla}_a \xi_b &= \xi_{\mathrm{a}} F^{\mathrm{a}}{}_{ba} \xi^a \\ &= \xi_{\mathrm{a}} h^{\mathrm{a}}{}_c \bar{T}^c{}_{ba} \xi^a \\ &= \xi^c \bar{T}_{cba} \xi^a. \end{aligned} \tag{4.2.21}$$

This equation predicts torsion-dependent deviations away from the geodesics of the Weitzenböck connection.

Field Equations

I turn finally to the field equations of TPG—i.e. the analogues of the Einstein equation (2.1.1) in the theory. These equations—derivable from an action principle[67]—read: (de Andrade et al., 2000, p. 4)

$$\partial_\sigma \left(h S_{\mathrm{a}}{}^{\sigma\rho} \right) - 4\pi \left(h j_{\mathrm{a}}{}^{\rho} \right) = 0, \tag{4.2.22}$$

with[68]

$$h = \det \left(h^{\mathrm{a}}{}_{\mu} \right), \tag{4.2.23}$$

$$h j_{\mathrm{a}}{}^{\rho} = -\frac{1}{4\pi} h h_{\mathrm{a}}{}^{\lambda} S_{\mu}{}^{\nu\rho} \bar{T}^{\mu}{}_{\nu\lambda} + h_{\mathrm{a}}{}^{\rho} \mathscr{L}_G, \tag{4.2.24}$$

$$S^{\rho\mu\nu} = \frac{1}{2} \left(K^{\mu\nu\rho} - g^{\rho\nu} \bar{T}^{\sigma\mu}{}_{\sigma} + g^{\rho\mu} \bar{T}^{\sigma\nu}{}_{\sigma} \right), \tag{4.2.25}$$

$$\mathscr{L}_G = \frac{h}{16\pi} S^{\rho\mu\nu} \bar{T}_{\rho\mu\nu}. \tag{4.2.26}$$

Using the definition of the components of the Weitzenböck connection (4.2.5), we can write (4.2.22) as[69]

$$\partial_\sigma \left(h S_{\lambda}{}^{\sigma\rho} \right) - 4\pi \left(h t_{\lambda}{}^{\rho} \right) = 0, \tag{4.2.27}$$

[65] Itself associated to the Abelian gauge theory of electromagnetism.

[66] Having set $q = m = 1$, where q is the charge of the particle in question and m is its mass.

[67] The relevant gravitational Lagrangian in TPG is $\mathscr{L}_G$, defined at (4.2.26).

[68] $h j_{\mathrm{a}}{}^{\rho}$ is related to gravitational energy-momentum in TPG; for details, see (de Andrade et al., 2000, p. 4).

[69] $t_{\lambda}{}^{\rho}$ encodes gravitational energy-momentum in TPG; there are thus clear and interesting parallels with the gravitational energy-momentum pseudotensor in GR (for some philosophical discussion of this object, see (Hoefer, 2000; Pitts, 2010; Lam, 2011; Read, 2020)).

with

$$ht_\lambda{}^\rho = \frac{h}{4\pi}\bar{\Gamma}^\mu{}_{\nu\lambda}S_\mu{}^{\nu\rho} + \delta_\lambda{}^\rho \mathscr{L}_G. \tag{4.2.28}$$

This is equivalent to the vacuum field equation of GR, $R_{ab} = 0$ (de Andrade et al., 2000, p. 5). Analogously, in the presence of matter, the field equations of TPG become

$$\partial_\sigma \left(hS_\lambda{}^{\sigma\rho}\right) - 4\pi \left(ht_\lambda{}^\rho\right) = 4\pi \left(hT_\lambda{}^\rho\right). \tag{4.2.29}$$

This is equivalent to the Einstein equation for GR with matter, (2.1.1) (de Andrade et al., 2000, p. 6).

Inter-Theory Translations

The above in hand, we can prove geometrization and recovery theorems, to translate from TPG to GR and vice versa (respectively). I begin with the geometrization theorem.[70]

Theorem 3. (Geometrization Theorem, TPG) *Let $\bar{\mathscr{M}} = \langle M, h^a{}_a, \Phi\rangle$ be a model of TPG, for which the field equations (4.2.29) hold, and let $\bar{\nabla}$ be the unique Weitzenböck derivative associated to $\bar{\mathscr{M}}$. Then all of the following hold:*

1. *There exists a unique metric field g_{ab} and associated Levi-Civita derivative operator ∇ which can be constructed from $h^a{}_a$ via (4.2.2) and (4.2.10), respectively.*
2. *∇ is the unique derivative operator on M such that, for all timelike curves on M with four-velocity field ξ^a,*

$$\xi^b\nabla_b\xi^a = 0 \quad \Longleftrightarrow \quad \xi^a\bar{\nabla}_a\xi_b = \xi^c\bar{T}_{cba}\xi^a. \tag{4.2.30}$$

3. *The curvature field $R^a{}_{bcd}$ associated with ∇ satisfies the Einstein equation, (2.1.1).*

As a consequence, for each DPM $\bar{\mathscr{M}}$ of TPG, there exists a unique DPM $\mathscr{M}$ of GR.

Proof: (1) follows from (4.2.2) and (4.2.10). For (2), begin with the force law of GR, $\xi^a\nabla_a\xi^b = 0$. Expanding the Levi-Civita derivative in a coordinate basis, we have

[70] What follows is elementary. It is possible to address this issue at a much higher level of mathematical sophistication—see (Read and Teh, 2018).

$$\xi^\mu \partial_\mu \xi^\nu + \xi^\mu \Gamma^\nu{}_{\rho\mu} \xi^\rho = 0. \tag{4.2.31}$$

Now using (4.2.11), we can write

$$\xi^\mu \partial_\mu \xi^\nu + \xi^\mu \bar{\Gamma}^\nu{}_{\rho\mu} \xi^\rho - \xi^\mu K^\nu{}_{\rho\mu} \xi^\rho = 0, \tag{4.2.32}$$

i.e.

$$\xi^a \bar{\nabla}_a \xi^b = \xi^a K^b{}_{ca} \xi^c. \tag{4.2.33}$$

Using (4.2.12), we then have

$$\xi^a \bar{\nabla}_a \xi_b = \xi^c \bar{T}_{cba} \xi^a, \tag{4.2.34}$$

which is the desired result: the force law for TPG. The converse translation is also possible, so the result follows. For (3), recall from §4.2.2 that the dynamical equations of TPG (4.2.29) are equivalent to the Einstein equation (2.1.1), in terms of the curvature field $R^a{}_{bcd}$ associated to the Levi-Civita derivative ∇ (see e.g. (de Andrade et al., 2000; Aldrovandi and Pereira, 2013)). Thus, for any DPM of TPG, one can construct a unique Lorentizan metric field g_{ab} with associated Levi-Civita derivative, and those geometric objects so constructed will satisfy (2.1.1). Thus, for every DPM $\bar{\mathscr{M}}$ of TPG, a unique DPM $\mathscr{M}$ of GR can be constructed therefrom. □

This in hand, I turn now to the recovery theorem.

Theorem 4. (Recovery Theorem, TPG) *Let $\mathscr{M}$ be a model of GR satisfying the Einstein equation* (2.1.1), *and let ∇ be the unique Levi-Civita connection associated to $\mathscr{M}$. Then all of the following hold:*

1. *There exists a tetrad field, $h^\alpha{}_a$, and unique associated Weitzenböck derivative $\bar{\nabla}$.*
2. *$\bar{\nabla}_a$ is flat (i.e. $\bar{R}^a{}_{bcd} = 0$ for the associated Riemann tensor $\bar{R}^a{}_{bcd}$).*
3. *For all timelike curves on M with four-velocity field ξ^a,*

$$\xi^b \nabla_b \xi^a = 0 \quad \Longleftrightarrow \quad \xi^a \bar{\nabla}_a \xi_b = \xi^c \bar{T}_{cba} \xi^a. \tag{4.2.35}$$

4. *The tetrad field $h^\alpha{}_a$ satisfies the field equations* (4.2.29).

The pair $\langle h^\alpha{}_a, \bar{\nabla} \rangle$ is not unique. A second pair $\langle h'^\alpha{}_a, \bar{\nabla} \rangle$ will also satisfy the stated conditions iff $h'^\alpha{}_a = M^\alpha{}_b h^b{}_a$, where $M^\alpha{}_b$ is a Lorentz matrix (i.e. a matrix satisfying (4.2.7)*).*

As a consequence, for each DPM $\mathcal{M}$ of GR, there exists a class of DPMs $\bar{\mathcal{M}}$ of TPG, differing by Lorentz transformations of the orthonormal indices of the tetrad field.

Proof: For (1), a tetrad field $h^{\mathrm{a}}{}_a$ can be constructed from a Lorentzian metric g_{ab} via (4.2.4); an associated Weitzenböck derivative operator $\bar{\nabla}$ can then be constructed via (4.2.5). For (2), we need to prove that any such $\bar{\nabla}$ is flat, in the sense that the associated Riemann tensor vanishes. To do so, recall that in a coordinate basis, the Riemann tensor reads

$$R^{\mu}{}_{\nu\rho\sigma} = \partial_\rho \Gamma^{\mu}{}_{\nu\sigma} - \partial_\sigma \Gamma^{\mu}{}_{\nu\rho} + \Gamma^{\tau}{}_{\nu\sigma}\Gamma^{\mu}{}_{\tau\rho} - \Gamma^{\tau}{}_{\nu\rho}\Gamma^{\mu}{}_{\tau\sigma}. \tag{4.2.36}$$

Substituting in the definition of the components of the Weitzenböck derivative (4.2.5), we observe pairwise cancellation, yielding the result

$$\bar{R}^{a}{}_{bcd} = 0. \tag{4.2.37}$$

For (3), the result follows from the same calculations as those presented for (2) in Thm. 3. For (4), we again recall from §4.2.2 that the Einstein equation of GR (2.1.1) and the field equations of TPG (4.2.29) are inter-translatable, so given that $\mathcal{M}$ is a DPM of GR, the tetrad field $h^{\mathrm{a}}{}_a$ constructed therefrom will satisfy (4.2.29).

To prove non-uniqueness, first recall from §4.2.2 that if $h^{\mathrm{a}}{}_a$ satisfies (4.2.2), then so too does $h'^{\mathrm{a}}{}_a = M^{\mathrm{a}}{}_{\mathrm{b}} h^{\mathrm{b}}{}_a$. Thus, if a tetrad field $h^{\mathrm{a}}{}_a$ can be constructed from a Loretnzian metric field g_{ab}, then so also can a distinct tetrad field $h'^{\mathrm{a}}{}_a$. As with $h^{\mathrm{a}}{}_a$, we can construct a Weitzenböck derivative associated to $h'^{\mathrm{a}}{}_a$, so $h'^{\mathrm{a}}{}_a$ satisfies (1)—and we know from §4.2.2 that this is the *same* Weitzenböck derivative as that associated to $h^{\mathrm{a}}{}_a$ (so, in turn, the torsion tensors associated to $h^{\mathrm{a}}{}_a$ and $h'^{\mathrm{a}}{}_a$ are also identical). Thus, the pair $\langle h'^{\mathrm{a}}{}_a, \bar{\nabla}\rangle$ satisfies (2)–(3).[71] For (4), note that the field equations (4.2.29) are invariant under transformations of tetrad indices, so if $\langle h^{\mathrm{a}}{}_a, \bar{\nabla}\rangle$ satisfies (4), then so too does $\langle h'^{\mathrm{a}}{}_a, \bar{\nabla}\rangle$. Thus, $\langle h'^{\mathrm{a}}{}_a, \bar{\nabla}\rangle$ also satisfies (1)–(4).

Thus, for each DPM $\mathcal{M}$ of GR, there is a class of DPMs $\bar{\mathcal{M}}$ of TPG, differing by 'internal' Lorentz transformations of the tetrad field. □

4.2.3 Background Independence

Before assessing the background independence of TPG, first consider Knox's claim that, in this theory, "the old entities from GR appear to be waiting in the wings" (Knox, 2011, p. 273). Specifically, Knox claims that in order for the tetrad field $h^{\mathrm{a}}{}_a$

[71] The uniqueness of $\bar{\nabla}$ in the class of models of TPG associated to a given model of GR is in stark disanalogy with the non-uniqueness of the derivative operator in the class of models of NGT related to a given model of NCT. Cf. §4.1.3.

to be defined via (4.2.2), we must specify a metric field g_{ab}; as a result, g_{ab} should be considered to have ontological priority over $h^{\mathrm{a}}{}_{a}$ in TPG. Knox claims that this is further brought out in the following points:[72] (Knox, 2011, p. 273)

1. The minimal coupling prescription in TPG uses the so-called *teleparallel derivative operator*—a recasting of the Levi-Civita derivative operator—rather than the Weitzenböck derivative operator. As a result, all non-gravitational dynamical equations in TPG take their simplest form relative to the inertial frames picked out by the Levi-Civita connection. Moreover, freely falling bodies follow geodesics of the metric field g_{ab} associated to the tetrad field via (4.2.2).[73]
2. The tetrad field $h^{\mathrm{a}}{}_{a}$ is not determined uniquely in TPG; rather, there exists a gauge redundancy encapsulated in (4.2.8). As a result, it is tempting to think that the metric field g_{ab}—for which there is no such redundancy—should be taken as ontologically prior to the tetrad field.

In light of such results, Knox concludes that there is no underdetermination between GR and TPG; rather, the latter is simply a redescription of the former. Assuming that this is correct, then since GR is a paradigm case of a background independent theory, so too must be TPG, which is a mere reformulation of GR. One may, however, question such reasoning. For instance, while it is true that one *could* define a tetrad field from a metric field g_{ab} via (4.2.2), Knox has not demonstrated that the opposite direction is not possible—i.e. that one cannot define a metric field g_{ab} from a tetrad field $h^{\mathrm{a}}{}_{a}$. Indeed, such is possible, via (4.2.3). Note also that any verdict on the background independence (or otherwise) of TPG imported via considerations of GR *à la* Knox presupposes a certain functionalist-*cum*-structuralist attitude to which I do not commit myself in this book (recall §3.8); rather, I choose to evaluate theories *vis-à-vis* their background independence *on their own terms*, in the manner of the naïve realist.

Consider (1) and (2) more explicitly. On (1), the argument for theory *identification* on this basis is, to repeat, something which I wish to resist: a more cautious position would state that, while *read literally* GR and TPG are committed to different ontologies, the formalism of the former is more natural than that of the latter, in light of (1) and given the empirical equivalence of the two theories. On (2), it is unclear why the existence of such a gauge redundancy in the tetrad field means that the theory is better interpreted as a reformulation of GR—perhaps Knox has in mind, following Dasgupta (2016), an Occamist norm to excise symmetry-variant, undetectable structure from one's roster of ontological commitments; but if so, this

[72] I have modified Knox's (2) to circumvent the issues mentioned in footnote 63.
[73] For recent philosophical discussion of minimal coupling, see (Read et al., 2018).

deserves to be made explicit.[74] In short, then: understood in a flat-footedly realist sense, TPG and GR are *different theories*, with *different* ontological commitments; moreover, they should be assessed *on their own terms* with respect to qualities such as background independence.[75]

The above in mind, turn now in earnest to the background independence of TPG. Beginning this time with Def. 10 (in terms of variational principles), note that (4.2.29) *is* derivable from an action principle (with Lagrangian (4.2.26)); moreover, since this Lagrangian is defined in terms of $h^{\mathrm{a}}{}_{a}$ *only*, with the tetrad fields being subject to Hamilton's principle, the latter two conditions in Def. 10 appear to be satisfied, so that TPG does appear to be background independent, on Def. 10. For this reason, the first conjunct of Def. 15 (in terms of non-trivial corner charges) is also satisfied; moreover, given that GR satisfies Def. 15 (Freidel and Teh, 2022), and given that the GR action differs from the TPG action only by a boundary term, one expects that TPG will also satisfy Def. 15. This, indeed, is corroborated in e.g. the writings of Hammad et al. (2019), in which the corner charges for TPG are computed explicitly and found to be non-trivial.[76]

In fact, however, the situation here is more nuanced than this story suggests, for in presentations of the derivation of (4.2.29) from the TPG action, only the $B^{\mathrm{a}}{}_{a}$ part of $h^{\mathrm{a}}{}_{a}$ is varied (Aldrovandi and Pereira, 2013, p. 177). This might lead one to worry that $h^{\mathrm{a}}{}_{a}$ has a non-variational component, and therefore that Def. 10 and Def. 15 are violated. Though, on the standard presentation, this is correct, recall from §3.5 that one *could* also vary the x^{a} part of $h^{\mathrm{a}}{}_{a}$, and obtain the same equations of motion. If one takes this approach, then Def. 10 is indeed satisfied by TPG; *unless* one also endorses the further modification to Def. 10 suggested in §3.5.

How does TPG fare on Belot's conception of background independence? The first thing to state here is that it is natural to take the tetrad field $h^{\mathrm{a}}{}_{a}$ in a model of the theory $\langle M, h^{\mathrm{a}}{}_{a}, \Phi \rangle$ to constitute the 'geometrical structure' in the theory (cf. §3.6.1). Assuming this in what follows, then, via the TPG field equations (4.2.29), one might think (by analogy with GR) that every gauge inequivalent model of the theory corresponds to a distinct geometry $h^{\mathrm{a}}{}_{a}$,[77] so that there is a one-one correspondence between the geometrical and physical degrees of freedom in the theory, and the theory is fully background independent, as per Def. 12.

[74] If $\bar{\nabla}$ were to possess the gauge redundancy that Knox indicates, then there would arguably be room, by analogy with Knox's reasoning in the case of NGT and NCT (Knox, 2014), to maintain that the spacetime structure of GR is fundamental. However, we have seen in §4.2.2 (and footnote 63 therein) that such redundancy is *not* present, *pace* Knox.

[75] Cf. §§4.3.1, 5.3.4. This methodology also guides the approach to gravitational energy presented by Dürr and Read (2019), written in response to the more structuralist approach of Dewar and Weatherall (2018).

[76] In a little more detail: Freidel and Teh demonstrate that the GR Lagrangian including the Gibbons-Hawking-York boundary term satisfies their definition of background independence (Freidel and Teh, 2022, p. 290); such an action, however, has the same boundary behavior as the TPG action. For further discussion of such boundary terms and their philosophical significance, see (Wolf and Read, 2023).

[77] Modulo the analogue of the worries that arise for GR in this regard (cf. §3.6.4).

There is one technical obstacle to this appraisal of TPG as satisfying Def. 12. We saw in (4.2.8) that there is an 'internal' Lorentz symmetry of the tetrad field, so that if $\langle M, h^{\mathsf{a}}{}_{a}, \Phi \rangle$ is a model of TPG, then so too is $\langle M, h'^{\mathsf{a}}{}_{a}, \Phi \rangle$, where $h^{\mathsf{a}}{}_{a}$ and $h'^{\mathsf{a}}{}_{a}$ are related by a Lorentz transformation, as per (4.2.8). As in §4.1.5, there are four open interpretive options when faced with this multiplicity of models. First, one may regard $h^{\mathsf{a}}{}_{a}$ and $h'^{\mathsf{a}}{}_{a}$ as equivalent geometries, and their respective models as gauge equivalent (indeed, this is plausible, since $\bar{\Gamma}^{\rho}{}_{\mu\nu}$ and $\bar{T}^{\rho}{}_{\mu\nu}$ are *invariant* under such 'internal' Lorentz transformations). Second, one may regard $h^{\mathsf{a}}{}_{a}$ and $h'^{\mathsf{a}}{}_{a}$ as *inequivalent* geometries, and their respective models as not gauge equivalent. In either of these cases, TPG's satisfaction of Def. 12 is not threatened.

Alternatively, one may regard $h^{\mathsf{a}}{}_{a}$ and $h'^{\mathsf{a}}{}_{a}$ as equivalent geometries, but $\langle M, h^{\mathsf{a}}{}_{a}, \Phi \rangle$ and $\langle M, h'^{\mathsf{a}}{}_{a}, \Phi \rangle$ as gauge inequivalent; as in our discussion of NGT in §4.1.5, such a move does not appear defensible, for such models differ in $h^{\mathsf{a}}{}_{a}$ and $h'^{\mathsf{a}}{}_{a}$ only. Finally, one may regard $h^{\mathsf{a}}{}_{a}$ and $h'^{\mathsf{a}}{}_{a}$ as inequivalent geometries, yet consider their models to be gauge equivalent. Whether such a move is ultimately plausible[78] is questionable; however, if it *is* endorsed, then Belot's Def. 12 is not satisfied.

Finally, consider whether TPG satisfies the alternative conceptions of background independence encapsulated in Def. 5 (no absolute objects), Def. 6 (no formulation with fixed fields), and Def. 9 (no absolute fields). One may begin by asking the following question: What is the status of the metric field η_{ab} in (4.2.2)? Since η_{ab} is fixed to be $\mathrm{diag}(-1,1,1,1)$ in all models of TPG, it is appropriate to regard this object as a *confined object*.[79] Since it is *not* the case that all confined objects should be considered absolute objects (for then even GR would manifest a large range of absolute objects—see §3.3.2), on this interpretation, in spite of the existence of this fixed structure in TPG,[80] η_{ab} does not lead to violations of background independence, on Def. 5, Def. 6, or Def. 9.[81]

To close, consider Pitts' point that use of the tetrad formalism in any theory gives rise to the presence of an absolute object (Pitts, 2006, p. 365). Given both local Lorentz freedom (in the sense of (4.2.8)) and coordinate freedom, one can bring the $\mathsf{a} = 0$ component of the tetrad field, $h^{0}{}_{a}$, into the form $(1,0,0,0)$ in the neighborhood of any $p \in M$; moreover, there cannot be any spacetime region in any model such that $h^{0}{}_{a}$ vanishes. Thus, TPG (or, in the example upon which Pitts focuses (Pitts, 2006, p. 365), GR coupled to a spinor field using an orthonormal tetrad field) gives an example of Jones-Geroch vector field: nowhere vanishing, everywhere timelike, and diffeomorphic to a vector field with components $(1,0,0,0)$ in some coordinate system—and so in turn, an example of an

[78] Or indeed coherent, in light of footnote 34.

[79] Indeed, we saw in §3.5 that such was a viable strategy when approaching parameterized theories.

[80] Which, if we are to follow (Smolin, 2006, p. 204), should ultimately be excised from our physics—see §3.3.2.

[81] Though below I suggest that Def. 5 may be violated in TPG for different reasons.

Andersonian absolute object.[82] Note that this case poses a straightforward problem for Def. 5 (no absolute objects), does not threaten Def. 6 (no formulation in terms of fixed fields), and only threatens Def. 9 (no absolute fields) if one regards a particular *component* of the tetrad field as being 'directly specified' in the KPMs of TPG.

Faced with this example, Pitts (2006, 2012) demonstrates how, by moving to the alternative spinor formalism of Ogievetsky and Polubarinov (1965), 'irrelevant variables' may be excised, and this absolute object may be avoided. If such a move can be made in the context of TPG,[83] then this theory's having an absolute object in this regard is avoided, thereby lifting a possible roadblock to its satisfaction of Def. 5 (no absolute objects). If such a move cannot be made, however, then TPG does violate Def. 5, and potentially also violates Def. 9 (no absolute fields).

4.3 Kaluza-Klein Theories

In the foregoing, I assessed the background independence of NGT (§4.1.2), NCT (§4.1.3), and TPG (§4.2.2). Another classical theory worthy of consideration is *Kaluza-Klein theory* (KKT). I introduce KKT in §4.3.1 and §4.3.2, before apprais-ing its background independence in §4.3.3.

4.3.1 Kaluza-Klein Theory

Transformations

KKT—introduced by Kaluza (1921) and developed by Klein (1926)—is an attempt to unify gravity and electromagnetism by construction of a higher-dimensional theory.[84] As usually presented,[85] KPMs of this higher-dimensional theory are pairs $\langle M, \tilde{g}_{AB} \rangle$, where M is a five-dimensional differentiable manifold, and $\tilde{g}_{AB}$ is a Lorentizan metric field on M.[86] DPMs of KKT are picked out via the five-dimensional vacuum Einstein equation

[82] As Pitts states, "The tetrad-spinor example seems rather more serious a problem for definitions of absolute objects than the Jones-Geroch cosmological dust example was, because the spinor field is surely closer to being a fundamental field than is dust or any other perfect fluid." (Pitts, 2006, p. 365).

[83] Without the resulting theory being distinct from TPG.

[84] For a more historical approach to KKT in the philosophy literature, see (Lehmkuhl, 2009, §6.4). The caveats raised in footnote 5 regarding the 'geometrization' of gravity also apply in this case.

[85] The following is a fairly canonical, coordinate-dependent presentation of KKT. For an alternative, coordinate-independent formulation of the theory, and discussion of the relative conceptual merits of one presentation over the other, see §4.3.2.

[86] I use capital Roman indices for abstract five-dimensional indices, and capital Greek indices for five-dimensional indices in a coordinate basis; to denote a set of five-dimensional coordinates, I sometimes use the notation $\tilde{x} = (x, x^5)$, where x denotes the first four coordinates in $\tilde{x}$.

$$\tilde{R}_{AB} = 0. \tag{4.3.1}$$

Here, $\tilde{R}_{AB}$ is the Ricci tensor associated to $\tilde{g}_{AB}$. Upon this metric field, we impose in every coordinate basis (following Kaluza (1921)) the *cylinder condition*,[87]

$$\partial_5 \tilde{g}_{\Lambda\Sigma} = 0. \tag{4.3.2}$$

Imposition of this condition carries a number of important consequences (Ferrari et al., 1989, pp. 70–71). First, generically one may parameterize a metric field $\tilde{g}_{AB}$ which satisfies this condition as

$$\tilde{g}_{AB} = \begin{pmatrix} g_{ab}(x) + \Phi(x) A_a(x) A_b(x) & \Phi(x) A_a(x) \\ \Phi(x) A_b(x) & \Phi(x) \end{pmatrix}; \tag{4.3.3}$$

the important point to note here is that all terms in (4.3.3) depend upon x only (not x^5). Second, if (4.3.2) is to be satisfied in every frame, then the most general allowed coordinate transformations are

$$x^\mu \to y^\mu = f(x^\mu), \tag{4.3.4}$$
$$x^5 \to y^5 = x^5 + g(x^\mu). \tag{4.3.5}$$

To see how this latter result is obtained, first expand $\partial_\Lambda \tilde{g}_{\Sigma\Theta}$ in a new coordinate basis (indices of which are denoted by primes):

$$\frac{\partial}{\partial x^\Lambda} \tilde{g}_{\Sigma\Theta} = \frac{\partial^2 y^{\Sigma'}}{\partial x^\Lambda \partial x^\Sigma} \frac{\partial y^{\Theta'}}{\partial x^\Theta} \tilde{g}_{\Sigma'\Theta'} + \frac{\partial y^{\Sigma'}}{\partial x^\Sigma} \frac{\partial^2 y^{\Theta'}}{\partial x^\Lambda x^\Theta} \tilde{g}_{\Sigma'\Theta'} + \frac{\partial y^{\Lambda'}}{\partial x^\Lambda} \frac{\partial y^{\Sigma'}}{\partial x^\Sigma} \frac{\partial y^{\Theta'}}{\partial x^\Theta} \frac{\partial}{\partial y^{\Lambda'}} \tilde{g}_{\Sigma'\Theta'}. \tag{4.3.6}$$

Hence

$$\frac{\partial}{\partial x^5} \tilde{g}_{\Sigma\Theta} = \frac{\partial^2 y^{\Sigma'}}{\partial x^5 \partial x^\Sigma} \frac{\partial y^{\Theta'}}{\partial x^\Theta} \tilde{g}_{\Sigma'\Theta'} + \frac{\partial y^{\Sigma'}}{\partial x^\Sigma} \frac{\partial^2 y^{\Theta'}}{\partial x^5 x^\Theta} \tilde{g}_{\Sigma'\Theta'} + \frac{\partial y^{\Lambda'}}{\partial x^5} \frac{\partial y^{\Sigma'}}{\partial x^\Sigma} \frac{\partial y^{\Theta'}}{\partial x^\Theta} \frac{\partial}{\partial y^{\Lambda'}} \tilde{g}_{\Sigma'\Theta'}. \tag{4.3.7}$$

By the cylinder condition (4.3.2), the left hand side of (4.3.7) is required to vanish; as a result, we also expect the terms of the right hand side of (4.3.7) to vanish term-by-term (in the general situation, no cross-term cancellation is possible). First considering the third term, we accordingly require that

[87] One understands (4.3.2) both as a restriction on the allowed coordinate systems in KKT—in analogy with imposition of the unimodular coordinate condition $\sqrt{-g} = 1$ in UGR (cf. footnote 31 of §3.3.3)—and (when translated into coordinate-independent language: see §4.3.2 below) as a restriction on $\tilde{g}_{AB}$.

$$\frac{\partial y^{\mu}}{\partial x^{5}} \frac{\partial y^{\Sigma'}}{\partial x^{\Sigma}} \frac{\partial y^{\Theta'}}{\partial x^{\Theta}} \frac{\partial}{\partial y^{\mu}} \tilde{g}_{\Sigma'\Theta'} + \frac{\partial y^{5}}{\partial x^{5}} \frac{\partial y^{\Sigma'}}{\partial x^{\Sigma}} \frac{\partial y^{\Theta'}}{\partial x^{\Theta}} \frac{\partial}{\partial y^{5}} \tilde{g}_{\Sigma'\Theta'} \stackrel{!}{=} 0. \tag{4.3.8}$$

The second term here is also required to vanish by the cylinder condition (4.3.2). For the first term to vanish, we require that

$$\frac{\partial y^{\mu}}{\partial x^{5}} \stackrel{!}{=} 0 \qquad \Rightarrow \qquad y^{\mu} = f(x^{\mu}), \tag{4.3.9}$$

which is (4.3.4).

Now consider the first and second terms on the right hand side of (4.3.7)—in fact, by symmetry, it suffices to consider just one of these terms.

$$\frac{\partial^2 y^{\Theta'}}{\partial x^5 \partial x^{\mu}} \frac{\partial y^{\Sigma'}}{\partial x^{\nu}} \tilde{g}_{\Theta'\Sigma'} = \frac{\partial^2 y^5}{\partial x^5 \partial x^{\mu}} \frac{\partial y^{\sigma}}{\partial x^{\nu}} \tilde{g}_{5\sigma} + \frac{\partial^2 y^5}{\partial x^5 \partial x^{\mu}} \frac{\partial y^5}{\partial x^{\nu}} \tilde{g}_{55} = 0 \tag{4.3.10}$$

$$\frac{\partial^2 y^{\Theta'}}{\partial x^{5\,2}} \frac{\partial y^{\Sigma'}}{\partial x^{\nu}} \tilde{g}_{\Theta'\Sigma'} = \frac{\partial^2 y^5}{\partial x^{5\,2}} \frac{\partial y^{\sigma}}{\partial x^{\nu}} \tilde{g}_{5\sigma} + \frac{\partial^2 y^5}{\partial x^{5\,2}} \frac{\partial y^5}{\partial x^{\nu}} \tilde{g}_{55} = 0 \tag{4.3.11}$$

(Setting $\Sigma = \Theta = 5$ does not yield any new, independent conditions.) From (4.3.11), we extract the condition

$$\frac{\partial^2 y^5}{\partial x^{5\,2}} \stackrel{!}{=} 0 \qquad \Rightarrow \qquad y^5 = \kappa\left(x^{\lambda}\right) x^5 + g\left(x^{\lambda}\right), \tag{4.3.12}$$

while from (4.3.10) we extract the condition

$$\frac{\partial^2 y^5}{\partial x^5 \partial x^{\mu}} = \frac{\partial}{\partial x^{\mu}} \left(\frac{\partial y^5}{\partial x^5} \right) = \frac{\partial \kappa\left(x^{\lambda}\right)}{\partial x^{\mu}} \stackrel{!}{=} 0, \tag{4.3.13}$$

from which we infer $\kappa\left(x^{\lambda}\right) = \text{const}$; finally, rescaling (4.3.12) and absorbing the factor of κ^{-1} into $g\left(x^{\lambda}\right)$ yields (4.3.5).

Field Equations

Thus far, KKT is equivalent to five-dimensional vacuum GR, with one auxiliary condition, (4.3.2). Given the form of the metric (4.3.3) which results from this condition, one must ask what the field equations (4.3.1) of the theory become, when (4.3.3) is substituted thereinto. If one were to do so at this point, one would obtain three dynamical equations which feature non-trivial couplings between g_{ab}, A^a and Φ.[88] In KKT, however, the goal is to obtain dynamical equations which represent coupled Einstein-Maxwell gravity *only*. In order to attain this result, one

[88] See e.g. (Morgado Patrício, 2013, p. 1).

imposes at this stage a second constraint: $\Phi = 1$, thus fixing this degree of freedom in the theory (Pasini, 1988, p. 293).[89] Then, rescaling $A^a \to 4\sqrt{\pi}A^a$, the (a,b)-components of (4.3.1) become

$$G_{ab} = 8\pi T_{ab}^{EM}, \tag{4.3.14}$$

with G_{ab} the 'Einstein tensor' associated to the g_{ab} components of $\tilde{g}_{AB}$, and

$$T_{ab}^{EM} = \frac{1}{4} g_{ab} F_{cd} F^{cd} - F_a{}^c F_{bc} \tag{4.3.15}$$

the stress-energy tensor of electromagnetism. The $(a,5)$-components of (4.3.1) become

$$F_{ab}{}^{;a} = 0, \tag{4.3.16}$$

with $F_{ab} = 2A_{[b;a]}$ the Faraday tensor. Thus, the field equations (4.3.1) of KKT amount to the Einstein equation of GR, plus the source-free Maxwell equations, coupled together.

Diffeomorphism Invariance

Consider the transformation of the $\tilde{g}_{\mu 5}$ piece of $\tilde{g}_{AB}$ (in some arbitrary coordinate basis) from one coordinate basis to another, subject to (4.3.2). We have, from (4.3.4) and (4.3.5),

$$\begin{aligned} \tilde{g}_{\mu 5} &\to \tilde{g}_{\mu 5} \frac{\partial y^5}{\partial x^5} + \tilde{g}_{55} \frac{\partial y^5}{\partial x^\mu} \frac{\partial y^5}{\partial x^5} \\ &= \tilde{g}_{\mu 5} + \tilde{g}_{55} \partial_\mu g\left(x^\lambda\right), \end{aligned} \tag{4.3.17}$$

so that

$$A_\mu \to A_\mu + \partial_\mu g\left(x^\lambda\right). \tag{4.3.18}$$

(4.3.18) is the transformation law for the electromagnetic vector potential under a $U(1)$ rotation. Since this result is a consequence of (4.3.5), it is sometimes written that (4.3.2) breaks the invariance group of diffeomorphisms of KKT from $\mathrm{Diff}(M)$ to a subgroup $\mathrm{Diff}(N) \times U(1)$, where N is a four-dimensional manifold; the

[89] Two points deserve to be made on this restriction. First, like (4.3.2), setting $\Phi = 1$ can (in the coordinate-dependent language in which KKT has been articulated up to this point) be understood both as a coordinate condition, and as a certain physical restriction on g_{AB}. Second, like (4.3.2), this condition can be understood as a field equation, thereby restricting the space of DPMs of KKT, or as a direct restriction on the space of kinematical possibilities.

Diff(N) invariance is identified with transformations (4.3.4); the $U(1)$ invariance is identified with transformations (4.3.5).

In this book, I avoid speaking in these terms, for three reasons. First, it is incorrect to state that (4.3.18) follows directly from (4.3.5)—rather, it follows as a result of (4.3.4) and (4.3.5) *together*. Thus, it is not clear that one should identify (4.3.5) with the $U(1)$ invariance encapsulated in (4.3.18). Second, the above reasoning leaves ill-specified at a mathematical level why (4.3.5) itself corresponds to $U(1)$ transformations. Third, the manifold associated to the Lie group $U(1)$ is S^1, the sphere. Speaking of $U(1)$ transformations at this point is apt to yield confusion with Klein's *compactification conditions*, introduced by Klein (1926), in which the x^5 coordinate is rendered periodic (sometimes, authors speak of 'compactification on S^1').[90] Note, however, that no such compactification conditions have yet been introduced.

For my purposes, the relevant observation is that (4.3.2) restricts the class of diffeomorphisms under which the theory is invariant to (4.3.4) and (4.3.5). Here, however, an interesting subtlety arises. Treating (4.3.2) as a field equation, if $\langle M, \tilde{g}_{AB} \rangle$ is a DPM of the theory, then $\langle M, d^* \tilde{g}_{AB} \rangle$ generically will not be, for arbitrary $d \in \text{Diff}(M)$. If that is correct, then KKT is *not* diffeomorphism invariant, on Def. 3.[91, 92] This I take to be the most natural understanding of the diffeomorphism invariance of KKT—although it is worth at this point highlighting a different possible approach here: on which one understands (4.3.2) to be picking out (4.3.4) and (4.3.5) as the allowed diffeomorphisms, *for which the theory is defined*. In that case, the theory is not even defined for any diffeomorphism which does take such a form—and the theory *will* be invariant under any diffeomorphism of the form (4.3.4) and (4.3.5). On this understanding, coordinate conditions such as (4.3.2) render theories *not generally covariant* (since the theory is only defined with respect to a privileged class of coordinate systems), but (on this construal) *diffeomorphism invariant*. (I return to this issue in §4.3.2.) In any case, I close this subsection by noting that $\tilde{g}_{55}$ is *invariant* under (4.3.4) and (4.3.5).

Justifying the Cylinder Condition

When Kaluza (1921) introduced the above theory, the cylinder condition (4.3.2) was not given a physical motivation. In his later development of the theory, Klein (1926) attempted to rectify this situation, arguing that such an explanation can be found if the x^5 coordinate is rendered *periodic*.[93] In so doing, we declare that

[90] These conditions are discussed below.

[91] One alternative is to treat (4.3.2) as restricting the *KPMs* of KKT (cf. §§3.5, 5.3.4). On this approach, remains the case that KKT is not diffeomorphism invariant, for if $\langle M, \tilde{g}_{AB} \rangle$ is a DPM of the theory, then on this understanding $\langle M, d^* \tilde{g}_{AB} \rangle$ will generically not even be a KPM.

[92] This is a verdict also endorsed by, for example, Vassallo (2015)—although in that case on the inference that the diffeomorphism group of KKT is Diff$(N) \times U(1)$, which has already been criticized.

[93] Sometimes, this condition is referred to as *compactification on* S^1; I avoid such terminology. Cf. §4.3.1.

points along x^5 which differ by a quantity $2\pi R_{\text{KK}}$ (where R_{KK} is the hitherto-undetermined *Kaluza-Klein radius*) are identical: (Rotger, 2014, p. 2)

$$x^5 \sim x^5 + 2\pi R_{\text{KK}}. \tag{4.3.19}$$

Now, for illustration, consider a massless real scalar field in five dimensions, $\phi\left(x, x^5\right)$. For such a field, we make the following Fourier expansion on the periodic x^5 dimension (Overduin and Wesson, 1997, §4.1)

$$\phi\left(x, x^5\right) = \sum_{n=-\infty}^{\infty} \phi^n(x)\, e^{inx^5/R_{\text{KK}}}. \tag{4.3.20}$$

Then, the associated equation of motion—here the Klein-Gordon equation—reads

$$\left(\partial^\mu \partial_\mu - \frac{\partial}{\partial x^{5\,2}}\right) \phi\left(x, x^5\right) = 0, \tag{4.3.21}$$

which yields the equations of motion for the Fourier components $\phi^n(x)$

$$\left(\partial^\mu \partial_\mu + m_n^2\right) \phi^n(x) = 0, \tag{4.3.22}$$

with

$$m_n = \frac{n}{R_{\text{KK}}}. \tag{4.3.23}$$

From (4.3.22) and (4.3.23), we observe that the only massless field is $\phi^0(x)$; the other fields $\phi^n(x)$ have masses of order R_{KK}^{-1}. If R_{KK} is small, then these masses are large, so in the low-energy limit, the only non-negligible Fourier component of $\phi\left(x, x^5\right)$ is $\phi^0(x)$, which is independent of x^5. In turn, this renders the physics of the $\phi\left(x, x^5\right)$ field in the five-dimensional theory effectively independent of the x^5 coordinate. Now, what holds in the above simple case of a real scalar field also applies for other fields in KKT, including $\tilde{g}_{AB}$. Once again, due to the periodicity of x^5, we can expand the components of $\tilde{g}_{AB}$ as a Fourier series,

$$\tilde{g}_{AB}\left(x, x^5\right) = \sum_{n=-\infty}^{\infty} g_{AB}^n(x)\, e^{inx^5/R_{\text{KK}}}, \tag{4.3.24}$$

whereby analogous reasoning applies (Overduin and Wesson, 1997, §4.1), and the only non-negligible component of $\tilde{g}_{AB}\left(x, x^5\right)$ is $\tilde{g}_{AB}^0(x)$, which is independent of the x^5 coordinate. In this way, rendering the x^5 coordinate periodic with small R_{KK} allows for a physical explanation of (4.3.2)—though, of course, such periodicity itself remains unexplained.

Variational Principles

With the above in mind, consider now how the equations of motion (4.3.14), (4.3.15), and (4.3.16), in turn derived from the KKT equation of motion (4.3.1), are obtainable from an action principle. We begin with the five-dimensional Einstein-Hilbert action,

$$S_{\mathrm{KK}} = \frac{1}{16\pi} \int d^5 x \sqrt{-\tilde{g}} R(\tilde{g}), \tag{4.3.25}$$

where here $\tilde{g} = \det(\tilde{g}_{AB})$, and $R(\tilde{g}) = \tilde{R}$ is the Ricci scalar associated to $\tilde{g}_{AB}$. Substituting in the specific form of the metric (4.3.3), upon which (4.3.2) has been imposed, we obtain, up to an extra integral over the x^5 coordinate, the standard action for coupled Einstein-Maxwell gravity,[94]

$$S_{\mathrm{KK}} = \frac{1}{16\pi} \int dx^5 \int d^4 x \sqrt{-g} \left(R(g) + \frac{1}{4} F_{ab} F^{ab} \right). \tag{4.3.26}$$

At this stage, however, one might think that we have a problem, for the integral over the x^5 coordinate will in general diverge. However, if we again impose that this coordinate is periodic, with (by convention) period $2\pi R_{\mathrm{KK}}$, we can overcome this problem, for in this case the integral becomes a constant. In that case, performing the x^5 integral in (4.3.26), we obtain (Bailin and Love, 1987, p. 1093)

$$S_{\mathrm{KK}} = \frac{2\pi R_{\mathrm{KK}}}{16\pi} \int d^4 x \sqrt{-g} \left(R(g) + \frac{1}{4} F_{ab} F^{ab} \right), \tag{4.3.27}$$

where R_{KK} is the radius of the periodic x^5 dimension. Hence, we see that, in addition to providing an explanation for (4.3.2), imposing periodicity of the x^5 coordinate allows one to avoid divergences in the derivation of the equations of motion (4.3.14), (4.3.15), and (4.3.16) from a five-dimensional action principle.

4.3.2 Coordinate-Independent Kaluza-Klein Theory

Thus far, I have presented KKT in a coordinate-dependent manner, through the imposition of the coordinate condition (4.3.2). The resulting theory, on the most straightforward reading, is neither generally covariant nor diffeomorphism invariant—although, as we have seen, there is an alternative understanding of

[94] One might worry that a shift has occurred here, from coordinate-dependent to coordinate-independent reasoning. This is correct, but unproblematic: the right way to understand (4.3.26) is as a coordinate-independent expression invoking objects whose definition is dependent upon conditions given in a coordinate-dependent manner. Note, in addition, that (4.3.26) and (4.3.27) continue to hold in the coordinate-independent presentation of KKT of §4.3.2.

the latter notion on which KKT *is* diffeomorphism invariant. In this subsection, I consider whether the story changes on moving to a different, less commonly-presented, *coordinate-independent* version of KKT. As before, KPMs are pairs $\langle M, \tilde{g}_{AB} \rangle$.[95] In this case, however, DPMs are picked out via (4.3.1),

$$\mathscr{L}_X \tilde{g}_{AB} = 0, \tag{4.3.28}$$

and *no* coordinate conditions.[96] (4.3.28) requires that the Lie derivative[97] of $\tilde{g}_{AB}$ along the integral curves of some Killing vector field X^A vanish[98]—in which case, $\tilde{g}_{AB}$ has certain *isometries*. Note that the X^A is *not* to be regarded as fundamental; rather, it is simply used to pick out—in a coordinate-independent manner—a direction along which $\tilde{g}_{AB}$ is constant.

How does this formulation of KKT reduce to that presented previously? The answer is straightforward: by aligning the x^5 coordinate axis along the direction of the X^A field, (4.3.28) reduces, in such a coordinate basis, to (4.3.2).[99] In an important sense, however, this latter version of KKT is more general than that presented above—for whereas in this version the theory may be described using any coordinate basis, in the original version of KKT, our presentation of the theory was restricted to those coordinate systems in which (4.3.2) held. Thus, this presentation of KKT renders the theory generally covariant.

What should be made of this generally covariant version of KKT *vis-à-vis* diffeomorphism invariance? Since X^A picks out a direction along which $\tilde{g}_{AB}$ is constant, if one begins with a model of this theory $\langle M, \tilde{g}_{AB} \rangle$, and then considers a model $\langle M, d^* \tilde{g}_{AB} \rangle$ in which $\tilde{g}_{AB}$ has been acted upon by a diffeomorphism d, one should *also* act upon X^A with d (in order to continue to identify the direction along which g_{AB} is constant); in other words, X^A should *not* be treated as a fixed field. In that case, (4.3.28) is invariant under all diffeomorphisms (since it contains no fixed fields, meaning that the notions of general covariance and diffeomorphism invariance do not come apart), so meaning that, in *this* presentation of the theory, KKT *is* diffeomorphism invariant. The moral is, then, the following: one can restore the diffeomorphism invariance of a theory by elevating coordinate conditions to coordinate-independent expressions not featuring fixed fields.[100]

It is worth dwelling a little longer on the question of the diffeomorphism invariance of KKT—is there another way of understanding the diffeomorphism

[95] The content of this subsection is elementary compared to that of (Gomes and Gryb, 2021), in which is presented a much more general and geometrical approach to Kaluza-Klein theory.

[96] One should also recall the dynamical equation $\Phi = 1$. In principle, one could write down a coordinate-independent version of this requirement, since it serves only to restrict certain degrees of freedom of $\tilde{g}_{AB}$.

[97] See (Wald, 1984, §C.2). [98] See (Wald, 1984, §C.3). [99] See e.g. (Wald, 1984, p. 440).

[100] If one *were* to treat X^A as a fixed field, then (4.3.28) would *not* be diffeomorphism invariant. To see this, one can take Killing's equation $\nabla_{[A} X_{B]} = 0$ and apply a diffeomorphism, treating X^A as a fixed field (here, $X_A := g_{AB} X^B$).

invariance of this theory? To this end, recall Dewar's distinction between *Penrose* and *Earman-Friedman* approaches to NGT (Dewar, 2015, p. 324).[101] While the Earman-Friedman approach is that presented in §4.1.3, the Penrose approach proceeds differently: in this case, a classical spacetime is defined as the direct product of a three-dimensional manifold A and a one-dimensional manifold T; on such manifolds are defined separate metric fields (Dewar, 2015, p. 318). Proceeding in this manner, it is natural to investigate the group $\mathrm{Diff}(A \times T)$, for if this is the same as the group of diffeomorphisms of NGT or NCT on the Earman-Friedman approach, then it seems that the *most general* class of possible diffeomorphisms in the Penrose approach coincides with the diffeomorphisms under which the theory is invariant.

In principle, one can make a similar move in KKT, in which the proponent of the analogue of the Penrose approach to NGT or NCT begins by defining metric fields on the product manifold $N \times S^1$, before then investigating the group $\mathrm{Diff}\left(N \times S^1\right)$. If this group of diffeomorphisms coincides with the group of diffeomorphisms under which KKT (in its original formulation presented above) is invariant (i.e. the group of smooth coordinate transformations which take DPMs of KKT to DPMs—(4.3.4) and (4.3.5)), then one has (a) rendered the theory diffeomorphism invariant, while (b) retaining a restriction to a certain allowed class of diffeomorphisms.[102]

Unfortunately, such an approach does *not* go through. First, for two manifolds M_1 and M_2, in general $\mathrm{Diff}(M_1 \times M_2) \ncong \mathrm{Diff}(M_1) \times \mathrm{Diff}(M_2)$. Second, even supposing that the diffeomorphism group of KKT *is* $\mathrm{Diff}(N) \times \mathrm{Diff}\left(S^1\right)$ (which is *not* the case), so that $\mathrm{Diff}(N)$ is identified with the transformations (4.3.4), the group $\mathrm{Diff}\left(S^1\right)$, as discussed in §4.3.1, is non-trivial, and not straightforwardly identifiable with the transformations (4.3.5).[103] Thus, the Penrose formulation of KKT does not provide any obvious means of stating that this theory is diffeomorphism invariant, yet has allowed transformations (4.3.4) and (4.3.5).[104]

4.3.3 Background Independence

Notwithstanding the fact that we have seen two different presentations of KKT—one on which it is neither generally covariant nor diffeomorphism invariant (at least on a natural understanding of the latter notion), and one on which

[101] The former since it is Penrose's choice of presentation of NGT (Penrose, 2004, ch. 17); the latter since it is the preferred presentation of Friedman (1983) and Earman (1989).

[102] Thus, such a strategy would afford a means of underwriting the alternative approach to the diffeomorphism invariance of KKT presented in §4.3.1, according to which this theory *is* diffeomorphism invariant.

[103] See e.g. (Khesin and Wendt, 2009, ch. 2).

[104] The analogue of these results also applies to the Penrose formulation of NGT.

it is both generally covariant and diffeomorphism invariant—the verdicts on the background independence of this theory are, in fact, independent of the presentation that one chooses.[105] To see this, first consider Def. 6 (in terms of fixed fields): since KKT does not contain any fixed fields, this definition is satisfied;[106] likewise, Def. 9 (in terms of absolute fields) is also satisfied. By contrast, Def. 5 (in terms of absolute objects) is more complex. On the one hand, the component $\tilde{g}_{55}$ is invariant under (4.3.4) and (4.3.5). However, at this point this object is still formally *unfixed*—meaning that the space of DPMs of KKT is partitioned into domains, points of which are related by allowed diffeomorphisms, and in each of which $\tilde{g}_{55}$ takes a different value. This situation thus resembles Torretti's counterexample to the equation of the absence of absolute objects and background independence (§3.3.3), and therefore—one might argue—does not present a violation of Def. 5. In fact though, recall that, as Pitts points out, such a case *does* involve an absolute object—the square root of the metric determinant (Pitts, 2006, pp. 17–18). In addition, recall that in KKT, $\tilde{g}_{55}$ is ultimately fixed via $\Phi = 1$. This extra condition restricts the DPMs of KKT to only one of the above-mentioned equivalence classes—in which case there again arises an absolute object in this example.

Consider next Def. 10 (in terms of variational principles). Recalling the KKT action (4.3.25), it is clear that all dependent variables therein (namely the one field, $\tilde{g}_{AB}$) are subject to Hamilton's principle, and represent physical fields. Thus, it appears that KKT qualifies as background independent, on Def. 10. One subtlety in this regard, however, is the following. Above, I suggested that the conditions (4.3.2) and $\Phi = 1$ be imposed as *field equations*—i.e. at the level of DPMs. But if this is the case, one should be able to *derive* these equations from an action principle—and it is not obvious that this can be achieved. In fact, only in the case in which these conditions are imposed at the level of *KPMs* does KKT satisfy uncontroversially Def. 10 (cf. §§3.5, 5.3.4).

A similar point can be made regarding Def. 15 (in terms of the existence of non-trivial corner charges). But there are also additional complications here. By stipulation at the level of kinematics, (the coordinate-independent formulation of) KKT imposes via (4.3.28) that the metric field g_{AB} have an isometry. But in that case, one is dealing with a *rigid* symmetry, for which there will not be associated non-trivial corner charges (essentially, one is in the domain of Noether's first rather than second theorem in this case).[107] But this means that the second clause of Def. 15 is violated, so KKT (or any other theory in which the metric field by

[105] This, thus, further illustrates that the connection between (say) diffeomorphism invariance and background independence is delicate: see §3.2.4.

[106] Omitting here the unnatural decision to regard X^A as being a fixed field—see §4.3.2.

[107] For an isometry, it needn't be the case that the associated corner charge *vanishes*, but it will be non-dynamical, and in that sense 'trivial'. My thanks to Luca Ciambelli for discussion on this point. (For further details on corner charges in particular contexts, see (Ciambelli, 2022).)

stipulation possesses isometries) does *not* qualify as background independent, according to Def. 15. (That Def. 15 captures that metric isometries are associated with 'background structure' is reasonable, and aligns with the selfsame suggestion made by Lam (2011).)

Finally, turn to Belot's account of background independence. Here, since $\tilde{g}_{AB}$ is the only fundamental geometric object in the theory, one should take this object to represent the geometrical structure (see §3.6.1) in the theory; then, any such metric fields related by the allowed transformations (4.3.4) and (4.3.5) can be considered *equivalent geometries* (as with GR). Moreover, as in the case of GR, models which differ by metric fields related by such transformations can be considered physically equivalent. In this case, KKT satisfies Def. 12, of *full* background independence.

I close with the following comment. As with Knox on TPG and GR (cf. §4.2.3), one might argue—based upon the fact that one can obtain the Einstein-Maxwell action (4.3.27) simply by integrating out the (periodic) x^5 coordinate in the Kaluza-Klein action (4.3.25)—that KKT is simply a reformulation of GR, rather than a distinct theory. If so, then, again, KKT should inherit the background independence of GR.[108] As before, however, for the purposes of this book I wish to resist this reasoning: KKT is *prima facie* ontologically distinct from four-dimensional Einstein-Maxwell theory, and should be evaluated *on its own terms*.[109]

4.4 Machian Relationism

In this section, I turn to the so-called *Machian relationist* spacetime theories constructed by Barbour and his collaborators,[110] often claimed to constitute 'relationist' alternatives to NGT and GR.[111] There are two principal motivations for undertaking this study: (i) it has been claimed (Gryb, 2010; Barbour, 2015) that such theories are background independent—such claims are in need of critical scrutiny; (ii) it is once again illuminating (from the point of view of attaining greater understanding of the target theories) to pursue the question of the background independence of these theories for its own sake.

[108] Or vice versa—one would have to argue for one direction over the other. (Teh (2013) makes similar comments in the context of 'holographic dualities'—on which, see §5.3.)

[109] In this regard, I agree with Vassallo (2015).

[110] For technical introductions, see e.g. (Barbour, 2010; Mercati, 2018); for philosophical discussions, see e.g. (Earman, 1989; Pooley, 2002, 2013, 2015).

[111] Clearly, the term 'relationist' must be defined in this context. Perhaps the most appropriate definition of a 'relational' theory here is: one which does not presuppose more spacetime structure than that corresponding to the orbits of the *similarity group*. The notion of a similarity group is defined in §4.4.2. In this context, the best-known 'relationist' alternative to NGT is *Barbour-Bertotti theory* (Barbour and Bertotti, 1982) (or, better, the scale-invariant version of that theory—see (Barbour, 2003)); the 'relationist' alternative to GR is *shape dynamics* (Barbour, 2010; Mercati, 2018).

In §4.4.1, I discuss some well-known issues regarding spacetime structure in NGT, as a prolegomenon to recalling the procedure for the construction of Machian relationist theories in §4.4.2. In §4.4.3, I discuss the background independence of these theories, both according to the definitions presented in §3 and according to two new definitions of background independence, proposed by members of the Machian relationist school.

4.4.1 Newtonian Spacetime Structure

In §4.1, I broadly passed over ongoing debates regarding the most appropriate spacetime setting for NGT.[112] Before introducing Barbour's Machian program, however, more should be said on these matters. To that end, recall a hierarchy of classical (i.e. pre-relativistic) spacetime structures relevant to our purposes.[113] Beginning with the richest structure, *Galilean spacetime* (sometimes: *neo-Newtonian spacetime*) is the spacetime structure $\langle M, t_{ab}, h^{ab}, \nabla \rangle$ familiar from §4.1.1, consisting of a four-dimensional differentiable manifold M, temporal 'metric' field t_{ab} of signature $(1,0,0,0)$, spatial 'metric' field h^{ab} of signature $(0,1,1,1)$,[114] and compatible derivative operator ∇. If one excises the derivative operator ∇ from this structure, one arrives at *Leibnizian spacetime*, $\langle M, t_{ab}, h^{ab} \rangle$. If one further removes the t_{ab} field from this structure, one obtains *Machian spacetime*, $\langle M, h^{ab} \rangle$; and if one also replaces h^{ab} with a conformal equivalence class of spatial 'metrics' $\left[h^{ab}\right]$ (i.e. class of h^{ab} related by $h^{ab} \sim \lambda^2 h^{ab}$, for some scalar function λ), one arrives at what I here dub *conformal Machian spacetime*, $\langle M, \left[h^{ab}\right] \rangle$.[115]

Written in coordinate-independent language, the dynamical equations of NGT presented in §4.1.2 make reference to all and only the structures of Galilean spacetime, whence the standard claim that Galilean spacetime is the 'natural' setting for NGT.[116] According to Barbour, however, only spatial angles (corresponding to structure encoded in $\left[h^{ab}\right]$)—and in particular *not* (i) a temporal metric,

[112] But not entirely, for recall the discussion in §4.1.5 of Knox's claim that, in NGT, the gravitational potential φ should be regarded as being spatiotemporal *in addition* to the classical spacetime structure $\langle M, t_{ab}, h^{ab}, \nabla \rangle$ (Knox, 2014). Relatedly, Saunders (2013) argues that the most appropriate spacetime setting for NGT is *Maxwell spacetime*, in which (in the differential-geometric formalism) the derivative operator ∇ of Galilean spacetime (see below) is replaced with an equivalence class of such derivative operators, $[\nabla]$. For discussion attempting to reconcile Knox and Saunders, see (Weatherall, 2016c, 2018; Dewar, 2018; Teh, 2018; Wallace, 2020).

[113] Since they do not bear upon my discussion below, I omit mention of *Aristotelian spacetime*, *Newtonian spacetime*, and *Maxwell spacetime* (though on the latter, see footnote 112); for details, see (Earman, 2006, ch. 2).

[114] As discussed in §4.1.1, strictly neither t_{ab} nor h^{ab} satisfies the definition of a metric field, since they violate the non-degeneracy condition.

[115] For more on conformal Newtonian structures, see (Dewar and Read, 2020).

[116] Though authors such as Saunders (2013) and Knox (2014) demur—see footnote 112. Relatedly: one must take into account the force equation for test particles (4.1.11) in order to argue for the Galilean

or (ii) spatial distances—are amenable to direct empirical access.[117] Motivated by the desire to excise such allegedly non-empirical structures, Barbour (2010, 2015) seeks to formulate an alternative to NGT set in *merely* conformal Machian spacetime; i.e. a theory the dynamics of which adverts to only this (again, allegedly) empirically accessible structure.[118] In the following subsections, I outline the crucial steps underlying the construction of theories of this kind.[119]

4.4.2 Machian Relationism

Configuration Space

For a given physical system, define its *configuration space* $\mathscr{Q}$ to be the space of possible instantaneous states of that system.[120] As the system evolves, the point in $\mathscr{Q}$ representing the system's instantaneous state traces out a continuous curve. On this picture, KPMs are (monotonically rising) curves in the product space formed from $\mathscr{Q}$ and a one-dimensional space $\mathscr{S}$ which represents time.[121] DPMs are those curves allowed by the dynamics—for example, in Lagrangian mechanics, DPMs are those curves which extremize the action, which is a certain specified functional of such histories (Pooley, 2013, p. 558).

From $\mathscr{Q}$, define the *relative configuration space* $\mathscr{Q}_{\mathrm{RCS}}$ to be $\mathscr{Q}$ quotiented by rigid translations and rotations (such that two histories in $\mathscr{Q}$ related by such transformations correspond to the same history in $\mathscr{Q}_{\mathrm{RCS}}$), and *shape space* $\mathscr{Q}_{\mathrm{SS}}$ to be $\mathscr{Q}$ quotiented by rigid translations, rotations, and dilatations (changes of scale).[122] The union of the groups of rigid translations, rotations, and dilatations is known as the *similarity group*. Since, in NGT, the similarity group encodes structure corresponding to the derivative operator ∇ and spatial distances, dynamics formulated on $\mathscr{Q}_{\mathrm{SS}}$ will not appeal to this structure. Hence, an alternative to NGT with dynamics formulated on $\mathscr{Q}_{\mathrm{SS}}$ will be set in a classical spacetime structure $\langle M, t_{ab}, [h^{ab}] \rangle$. In order to arrive at a dynamics set in conformal Machian spacetime, it remains to eliminate the primitive temporal 'metric', t_{ab}.

Emergent Temporality

Initially, one might distinguish histories that correspond to a single curve in a configuration space (whether $\mathscr{Q}$, $\mathscr{Q}_{\mathrm{RCS}}$, or $\mathscr{Q}_{\mathrm{SS}}$) being traced out at different rates

invariance of NGT—for, as we have already seen, the Poisson equation (4.1.10) is invariant under the richer class of Leibniz transformations.

[117] For some critical evaluation of this claim, see (Pooley, 2013, §6.2).

[118] Note that Barbour does not himself formulate explicitly his Machian relationist theories in terms of Machian spacetime—rather, he presents them in terms of *configuration space* (see below).

[119] Here, I follow (Barbour, 2010; Pooley, 2013; Mercati, 2018).

[120] One may therefore, for a given $\mathscr{T}$, set $\mathscr{Q}$ in constrast with $\mathscr{K}$—cf. footnote 11 of §2.1.

[121] Such a curve may be identified with an element of $\mathscr{K}$—again, cf. footnote 11 of §2.1.

[122] For further detail, see e.g. (Barbour, 2010, §2).

with respect to the primitive temporal 'metric'. The Machian relationist, however, views each curve in configuration space as corresponding to exactly one possible history. Accordingly, they seek to eliminate $\mathscr{S}$, electing instead to work with a 'timeless' dynamics using $\mathscr{Q}$ alone. One way to do this is to first equip the configuration space with a metric, referred to as the *kinetic metric* (Pooley, 2013, p. 558), which is defined in terms of the spatial (conformal) 'metric'; in the particle mechanics case, this reads[123]

$$ds_{\text{kin}}^2 = \sum_{a=1}^{N} \frac{m_i}{2} d\vec{x}_i \cdot d\vec{x}_i, \tag{4.4.1}$$

where the sum is over the N particles in the system under configuration, and the m_i are the masses associated with these particles. Using the kinetic metric, DPMs are then picked out via a variational principle: $\delta S = 0$—continuing with the example of particle mechanics, this takes the specific form

$$S = \int \sqrt{F_E} ds_{\text{kin}}^2 = 2 \int d\lambda \sqrt{F_E T_{\text{kin}}}, \tag{4.4.2}$$

$$F_E := \left(E - V\left(x_1^i, \ldots, x_N^i\right)\right), \tag{4.4.3}$$

$$T_{\text{kin}} := \frac{1}{2} \sum_{a=1}^{N} m_a \frac{dx_a^i}{d\lambda} \cdot \frac{dx_a^i}{d\lambda}. \tag{4.4.4}$$

Solutions correspond to Newtonian histories of a system of N particles with a total energy E interacting according to a potential $V\left(x_1^i, \ldots, x_N^i\right)$. T_{kin} takes the form of the standard Newtonian kinetic energy, but λ represents an arbitrary parameterization of paths in $\mathscr{Q}$. The Euler-Lagrange equations of motion corresponding to (4.4.2) are

$$\frac{d}{d\lambda}\left(\sqrt{\frac{F_E}{T_{\text{kin}}}} m_a \frac{dx_a^i}{d\lambda}\right) = \sqrt{\frac{T_{\text{kin}}}{F_E}} \frac{\partial F_E}{\partial x_a^i}. \tag{4.4.5}$$

These simplify, reducing to the standard form of Newton's second law, when the freedom in the choice of λ is used to set $F_E = T_{\text{kin}}$, so that $E = T_{\text{kin}} + V$. As Pooley (2013, §6.2) puts it, "The Machian sees the equations as *defining* an emergent temporal metric in terms of the temporal parameter that simplifies the dynamics of the system as a whole." This procedure, of (i) defining a metric on configuration space,[124] (ii) using this to articulate a variational principle, and (iii) choosing the

[123] My presentation in this section follows closely that of (Pooley, 2013, §6.2).

[124] In the particle mechanics example used above, $\mathscr{Q}$ was used; however, in order to realize the Machian relationist ambitions, ultimately this should be performed on $\mathscr{Q}_{\text{SS}}$: this task can be achieved via *best matching* (for which, see below).

parameterization in the resulting equations of motion which maximally simplifies the dynamics, and treating this as the *definition* of a temporal parameter, is known in the literature on Machian relationism as *Jacobi's principle*: for original sources on this procedure, see e.g. (Barbour, 1994, 2008); for some recent philosophical discussion (related in particular to spacetime functionalism—cf. §3.8), see (Butterfield and Gomes, 2021a,b).

Best Matching

In order to apply Jacobi's principle on $\mathscr{Q}_{\text{SS}}$, one first needs (by analogy with the above) to define a metric on this space, and associated variational principle. The purpose of *best matching* is to achieve this task.[125] The procedure works as follows. Consider a class of paths in $\mathscr{Q}$, all corresponding to the same path in $\mathscr{Q}_{\text{SS}}$. A metric on $\mathscr{Q}$ will in general assign to each such path in $\mathscr{Q}$ a different length; as a result, the length which should be assigned to the path in $\mathscr{Q}_{\text{SS}}$ is thus far underdetermined. However, starting from any given point $p \in \mathscr{Q}$, one can use the action of the similarity group on $\mathscr{Q}$ to define a unique length, by shifting the points of any curve through p along the corresponding similarity group orbits so as to extremize the length assigned to the curve, in turn designated as the length of the curve in $\mathscr{Q}_{\text{SS}}$.[126, 127] Having assigned each curve through $\mathscr{Q}_{\text{SS}}$ a unique length, the variational prescription of §4.4.2—i.e. Jacobi's principle—may then be applied to eliminate primitive temporal structure.

Relativistic Machian Relationism

The best matching prescription can be applied not only to NGT, but also to relativistic theories, including GR. To see how this proceeds, first assume that instantaneous space has the topology of some closed 3-manifold without boundary, Σ. The natural analogue of $\mathscr{Q}$ is then $\text{Riem}(\Sigma)$, the space of Riemannian 3-metrics δ_{ij} on Σ.[128] The analogue of $\mathscr{Q}_{\text{RCS}}$ is *superspace*, the space of 3-*geometries*.[129] Two points $\langle \Sigma, \delta_{ij} \rangle$ and $\langle \Sigma, \delta'_{ij} \rangle$ of $\text{Riem}(\Sigma)$ are taken to correspond (at least for the purposes of the discussion here—although recall Belot's comments at (Belot, 2008, p. 201), quoted in §2.2) to the same 3-geometry just in case they are isomorphic, i.e. just in case for some diffeomorphism d of Σ, $\delta'_{ij} = d^* \delta_{ij}$. Superspace is thus $\text{Riem}(\Sigma)/\text{Diff}(\Sigma)$, the quotient of $\text{Riem}(\Sigma)$ by the group of diffeomorphisms of Σ. In turn, the analogue of $\mathscr{Q}_{\text{SS}}$ in this case is *conformal superspace*: the quotient of superspace by scale transformations (Pooley, 2013, p. 563). Having constructed

[125] See (Barbour, 1999, ch. 7) for an informal discussion, and (Pooley, 2013, §6.2) for a philosophical presentation.

[126] It can be helpful to think of $\mathscr{Q}$ as a principal fibre bundle over $\mathscr{Q}_{\text{SS}}$—see (Mercati, 2018, ch. 5).

[127] Best-matched theories are only empirically equivalent to NGT in certain regimes—see (Barbour, 2003, pp. 1556–7) and (Pooley, 2013, p. 561).

[128] δ_{ij} here should not be confused with the Kronecker delta.

[129] There is no connection here with 'superspace' in the sense associated with supersymmetry.

$\mathscr{Q}_{SS}$ in this way, one may again best match via a suitable action principle to obtain one's relational theory. The resulting such theory is called *shape dynamics*.[130]

Machian Relationist Spacetime Settings

A very natural question to ask of Machian relationist theories is this: what is their spacetime structure? If one wishes to retain a commitment to fundamentality of the topology of time, and to continue to present a four-dimensional spacetime setting for these theories, then one natural approach is to regard their spacetime setting as being (closely related to) conformal Machian spacetime, $\langle M, [h_{ab}] \rangle$, where $[h_{ab}]$ is a conformal equivalence class of four-dimensional 'metric' fields h_{ab} on the 4-manifold M.[131] There are, however, a number of notable differences between conformal Machian spacetime, in the form presented above in §4.4.1, and the spacetime structure of these Machian theories, which it will be important to clarify.

Begin with the Machian relationist theory of scale-invariant Newtonian particle dynamics (Barbour, 2003). In this case, the conformal transformations on each 3-geometry are *global*—i.e. they are independent of spatial position. One way to implement this in the language of differential geometry would be to insist that taking $[h^{ab}]$ to be composed of appropriate geometric objects related by conformal transformations $h^{ab} \sim \lambda^2 h^{ab}$, these transformations must be *spatially constant*—they must not vary on any given 3-geometry. This condition reads $h^{ab} d_b \lambda = 0$: d denotes the exterior derivative; thus, one does not need to specify a *particular* derivative operator on M in order to articulate this condition.[132] Thus, in this case, one takes the spacetime structure to be $\langle M, [h^{ab}] \rangle$, where (elements of) $[h^{ab}]$ are related by conformal transformations subject to the spatial constancy condition.

By contrast, in the case of shape dynamics (i.e. the Machian relationist alternative to GR), conformal transformations need not be spatially constant (on each 3-geometry, they are *local* conformal transformations—i.e. Weyl transformations), but (*unlike* the previous case), on each 3-geometry, there is a fact about the global volume of the universe (Mercati, 2018, p. 6). In the differential-geometric formalism, such a fact can be encoded in one additional degree of freedom Ξ; thus, in this case, one takes the spacetime structure to be $\langle M, [h^{ab}], \Xi \rangle$, where (elements of) $[h^{ab}]$ are related by conformal transformations (in this case not subject to the spatial constancy condition), and Ξ encodes facts about the global volume

[130] See e.g. (Barbour, 2010; Mercati, 2018) for details; technical complications involved in best matching in the relativistic case are discussed by Gomes (2011).

[131] If the Machian relationist rejects the view that time has topological structure, then, now written as (sets of) tuples of geometric objects, KPMs of Machian relationist theories such as shape dynamics are best understood as sets $\{ \langle \Sigma, [\delta_{ij}] \rangle \}$ of conformal (dynamical—not fixed) Riemannian 3-geometries.

[132] Note that this condition still allows the conformal factor to vary from 3-geometry to 3-geometry. Note also that, in specifying spatial constancy, one will have to choose a representative of the equivalence class $[h^{ab}]$.

of the universe. Note that in shape dynamics, Ξ need not be fixed—rather, it is a dynamical entity in itself. This will be relevant in our discussions of background independence below.

The foregoing amounts to the following. While, in scale-invariant particle dynamics, one can compare the conformal rescalings at different points in the same 3-geometry, there is no fixed fact about global volume. By contrast, in shape dynamics, one *cannot* compare conformal rescalings at different points in the same 3-geometry, but there *is* a global fact about the volume of the universe.[133]

4.4.3 Background Independence

The above presentation of the central aspects of Barbour's Machian relationist program is sufficient in order to undertake an assessment of the background independence of these theories. (Of course, I am eliding a vast amount of rich and fascinating dynamical detail: see (Mercati, 2018) for the *locus classicus* here.) In this subsection, I first discuss the background independence of the Machian relationist alternatives to Newtonian mechanics (in particular, scale-invariant particle dynamics); then, I do the same for shape dynamics. Finally, I discuss and assess new proposals for a definition of background independence, due to two prominent proponents of the Machian approach: Barbour (2010) and Gryb (2010).

Newtonian Machian Theories

Consider scale-invariant particle dynamics, with its spacetime structure $\langle M, [h^{ab}] \rangle$, as discussed in §4.4.2. In this case, each of the h^{ab} in the conformal equivalence class $[h^{ab}]$ is a fixed field; while considering only conformal structure eliminates facts about distances on 3-geometries encoded by this structure, facts about *angles* are retained, and continue to be fixed. Thus, there is a good case to be made that these theories, like their Newtonian counterparts, continue to contain fixed structures.

The upshot for the background independence (or otherwise) of these theories is, then, the following. Since $[h^{ab}]$ is a fixed field—*a fortiori* an absolute field, and an absolute object, the theories in question violate Def. 5 (no absolute objects), Def. 6 (no formulations in terms of fixed fields), and Def. 9 (no absolute fields). Moreover, this object is still present in the action principles for such theories: representatives of the class are used to perform index contraction in order to construct the kinetic metric, which is then used in an action principle (see e.g. (Pooley, 2013, p. 558), and §4.4.2 above). As a result, it is not the case that all the dependent variables in

[133] It is worth noting that there is as yet no Machian relationist reformulation of Newtoinan *field* theories—e.g. Newton-Cartan theory (see §4.1). Constructing such theories would be an interesting task for future pursuit.

the action principles of such theories are subject to Hamilton's principle—meaning that Def. 10 (variational principles) is also violated, as is Def. 15 (non-trivial corner charges).[134] Finally: the presence of the field structure $\left[h^{ab}\right]$ in the models of these theories, and no other (primitive) structure which could plausibly be regarded as being 'geometrical', means that Belot's definition of background independence Def. 12 is also violated, and these theories are classified as being *fully background dependent*.

Shape Dynamics

We have seen that the four-dimensional spacetime setting of shape dynamics is plausibly written $\langle M, [h_{ab}], \Xi \rangle$, where representatives of $[h_{ab}]$ are *dynamical* 'metric' fields of signature $(0, 1, 1, 1)$, related by *local* (and not necessarily spatially constant) conformal transformations; Ξ encodes facts about global volume. Since $[h_{ab}]$ is dynamical, such a field will not lead to violations of Def. 5 (no absolute objects), Def. 6 (no formulations in terms of fixed fields), or Def. 9 (no absolute fields). What of Ξ? Likewise, this object is dynamical in shape dynamics, and hence does not qualify as a fixed field, absolute field, or absolute object, so the respective definitions of background independence are not violated by this object.[135]

Does the square root of the 'metric' determinant (or at least, the determinant of a representative of the conformal class) count as an absolute object? The situation here is the same as that for NCT-II, already discussed above: since each $h^{ab} \in \left[h^{ab}\right]$ is *degenerate*, such an object is identically zero—but this need not impugn a verdict that background independence *à la* Def. 5 is violated, because zero remains a perfectly valid value for such an object. Taking this line, shape dynamics—unlike even its (scale-invariant) Newtonian particle mechanics counterpart—*does* satisfy Def. 6 and Def. 9 of background independence, but arguably still violates Def. 5.[136]

Turning to definitions of background independence in terms of variation principles, one first needs to be a little more explicit about the variational formulation of this theory. One way to proceed here is in terms of the action for the *linking theory* presented by Gomes and Koslowski (2012),[137, 138] which reads

[134] It's not even obvious that Def. 15 should be regarded as being applicable here, since it was formulated in the context of field theories.

[135] The situation regarding this object is thus not akin to Torretti's counterexample raised in the context of GR (cf. §3.3.3), for which it would have to be the case that within a given model, Ξ is fixed and non-dynamical, but this object may change from one model to the next.

[136] If one understands the kinematics of shape dynamics *à la* footnote 131, then one does not face this issue of degeneracy, for the determinants associated with each of the 3-metrics are not degenerate. Thus, on this formulation, the violation of Def. 5 is more evident.

[137] The linking theory is also discussed in (Mercati, 2018, ch. 12).

[138] One might consider it to be a roadblock that shape dynamics is usually written in terms of the Hamiltonian rather than the Lagrangian formalism (see e.g. (Mercati, 2018)—though for other discussions of Lagrangian formulations of shape dynamics, see (Gielen et al., 2018; Sloan, 2022)). However, there does not appear to be any substantial issue with reformulating Def. 10 in terms of a Hamiltonian. In any case, in what follows I'll mostly continue to use the Lagrangian formalism, and thus Def. 10 as stated.

$$S = \int dtd^3x \left(\frac{1}{4N} G^{abcd} \left(\dot{g}_{ab} - \mathscr{L}_\xi g_{ab} - \rho g_{ab} \right) \left(\dot{g}_{cd} - \mathscr{L}_\xi g_{cd} - \rho g_{cd} \right) + NT_\phi R \right), \tag{4.4.6}$$

where $G_{abcd} := g_{ac}g_{bd} - \frac{1}{2}g_{ab}g_{cd}$ is the 'supermetric', and ϕ and ρ are scalar fields; $\mathscr{L}_\xi$ denotes the Lie derivative along the integral curves of a vector field ξ^a; for the meaning of the operator T_ϕ (associated with canonical transformations), see (Gomes and Koslowski, 2012). When one fixes (4.4.6) such that $\phi = \phi_0$ and $-\rho = \dot{\phi}_0$, one arrives at the Lagrangian for shape dynamics. The important point to note though, having done so, is that all variables in the Lagrangian are subject to Hamilton's principle, and represent physical fields—accordingly, Def. 10 appears to be satisfied.

For Def. 15 (in terms of the existence of non-trivial corner charges), a little more work is required. In principle, one would need to vary the action (4.4.6) in order to compute the associated symplectic potential, from which one could then compute the corner charges for shape dynamics (by imposing the constraints associated with restricting the linking theory to that theory)—such variations would be in the manner of Gomes (2013). Thankfully, a shortcut here is available. In canonical GR, the charges $\mathscr{H}_\xi$ associated with an arbitrary diffeomorphism tangent to some spacelike Cauchy surface Σ with generating vector field ξ^a can be written in terms of a bulk constraint $\mathscr{H}_\xi^\Sigma$ and a corner charge $\mathscr{H}_\xi^S$, where

$$\mathscr{H}_\xi^\Sigma = -\int_\Sigma \xi_a \tilde{\nabla}_b \tilde{p}^{ab}, \tag{4.4.7}$$

$$\mathscr{H}_\xi^S = \int_S \sqrt{q} \tilde{s}_a n_b \nabla^a \xi^b, \tag{4.4.8}$$

where $\tilde{p}^{ab}$ is the momentum density conjugate to the induced metric $\tilde{g}_{ab}$ on Σ, $\tilde{\nabla}$ is the induced Levi-Civita connection on Σ, $\tilde{s}_a$ is a spacelike vector normal to S and n_a (which picks out a direction normal to Σ), and q is the determinant of the metric $q_{ab} = \tilde{g}_{ab} - \tilde{s}_a \tilde{s}_b$, which is the induced metric on the corner S of Σ (Freidel et al., 2020, p. 14). At this point, this formalism is common to both GR and shape dynamics; one specializes to one theory or the other by way of a different constraint surface. But in either case, on-shell, $\mathscr{H}_\xi^\Sigma$ vanishes, meaning that the charges associated with such diffeomorphisms are non-vanishing corner objects—suggesting that Def. 15 is satisfied by shape dynamics, at least for diffeomorphisms associated with transformations of the *spatial* coordinates.[139] For diffeomorphisms associated with transformations in

[139] So, strictly speaking, what's suggested at this point is that shape dynamics is *spatially* background independent, in the sense of §4.1.5. If one understands the basic objects of shape dynamics to be 3-geometries, then arguably there is nothing more to say here; on the other hand, if one takes a four-dimensional view on the basic spatiotemporal ontological commitments of this theory, then there

the timelike direction, we already know from §3.7—by writing the models of GR in the covariant rather than canonical formalism (which, indeed, is the more general approach anyway, since of course not all GR models admit of a $3+1$ decomposition)—that such transformations should also be associated with non-trivial conserved charges (this, indeed, is part of the lesson of (Freidel and Teh, 2022)). In the case of shape dynamics, however, the story is not so simple: in the canonical formalism, the constraints associated with these diffeomorphisms are complex, and there is no reason to expect that the associated charge should be a corner object on-shell (Lee and Wald, 1990). As a result, our tentative verdict—(as in several other cases by now) perhaps surprisingly—is that shape dynamics isn't obviously fully background independent, on Def. 15.[140, 141, 142]

Turning finally to Belot's approach to background independence, the situation is again somewhat subtle, albeit for different reasons. Consider a Machian reformulation of GR coupled to matter, as presented in e.g. (Barbour et al., 2002, pp. 17ff.) (see also (Barbour, 2010; Mercati, 2018)). In general, the dynamical equations of such theories are highly complex—see, for example, (Barbour et al., 2002, p. 18) for the dynamical equations of a Machian alternative to GR coupled to a real scalar field, and (Barbour et al., 2002, p. 21) for the dynamical equations of a Machian alternative to GR coupled to a vector field. Without significant calculational investment to map the solution space of these partial differential equations, it remains unclear whether each configuration of the geometrical fields (most naturally, $[h_{ab}]$) corresponds to a unique configuration of the non-geometrical (i.e. matter) fields. As with GR, however, it would be highly surprising if such were the case (recall §3.6.4 and the counterexamples to such a claim regarding GR); thus, on Belot's account, the most likely verdict regarding the theories in question is that they satisfy neither Def. 12 nor Def. 11—that is, are neither fully background independent nor fully background dependent.

A summary of these verdicts on the background independence of both classes of Machian relationist theories can be found in Table 4.1. The verdicts are both (a) interestingly different from one another, and (b) interestingly negative in the case of scale-invariant particle dynamics—from which one sees that even a thoroughgoing 'relationist' theory can violate numerous definitions of background independence.

remain open questions regarding the diffeomorphisms associated with *timelike* transformations—these questions I take up in the main text.

[140] Though, to repeat footnote 139, the verdict of background independence is more straightforward in the case of shape dynamics when 3-geometries are taken to be ontologically basic.

[141] Note that in my discussion here I haven't included consideration of e.g. Gibbons-Hawking-York boundary terms. For such terms in the context of shape dynamics, see (Gomes, 2013); for their relevance to Def. 15, see (Freidel and Teh, 2022).

[142] I'm grateful to Henrique Gomes for discussions on how to apply Def. 15 to shape dynamics.

Table 4.1 The appraisals delivered by Def. 5 (no absolute objects), Def. 6 (no formulation featuring fixed fields), Def. 9 (no absolute fields), Def. 10 (in terms of variational principles), Def. 12 (Belot's approach), and Def. 15 (in terms of the existence of non-trivial corner charges) on the background independence of NGT, NCT, NCT-II, TPG, KKT, SIPD (i.e. scale-invariant particle dynamics), and SD (i.e. shape dynamics). ✓ indicates that the theory is judged to be background independent; ✗ indicates that the theory is judged to be background dependent; and * indicates that there are subtleties or caveats regarding the verdict (for the exact nature of these, I refer the reader back to the main text).

	Def. 5	Def. 6	Def. 9	Def. 10	Def. 12	Def. 15
NGT	✗	✗	✗	✗	✗*	✗
NCT	✗	✗	✗	✗	✓*	✗
NCT-II	✗*	✓	✓	✓	✓*	✓*
GR	✗	✓	✓	✓	✓*	✓
TPG	✗	✓	✓	✓*	✓*	✓*
KKT	✗	✓	✓	✓*	✓*	✗*
SIPD	✗	✗	✗	✗	✗	✗
SD	✗*	✓	✓	✓	✓*	✓*

Best Matching and Background Independence

Finally, it should be noted that advocates of the Machian relationist agenda have proposed their own definitions of background independence. On the notion of background independence, Barbour writes:

> As for background independence... it should by now be clear that for me this means that nothing extraneous to shape space should play a role in dynamics.
>
> (Barbour, 2015, p. 60)

The thought is that any theory constructed in line with best matching and the elimination of primitive temporal metric structure should qualify as background independent; this may be elevated to the status of a definition as follows:

Definition 16. (Background independence, Machian version 1) *A theory is* background independent *iff its dynamics is formulated on shape space, and it does not possess primitive temporal metric structure in its formulation.*

On Def. 16, all Machian relationist theories count as background independent. One immediate worry regarding this definition, however, is the following: one might question whether it delivers the appropriate verdict on certain spacetime theories. For example, both NGT and its Machian alternative possess fixed structures in their formulations—and in this sense it is *prima facie* reasonable to judge both to be background dependent. This is contrary to Def. 16, according to which the latter is to be regarded as background independent. Moreover, it appears that, on Def. 16, GR counts as background *dependent*, for the dynamics of this theory is not formulated on shape space.[143]

A likely Machian response to the above worry regarding NGT would run as follows: 'We are concerned not (only) with the dynamical/non-dynamical distinction in physical theories, but with the excision of 'unobservable' structure. Insofar as any structure capable of generating Leibnizian "distinctions without a difference" is eliminated, our relationist goal has been realized—and the resulting theory may be judged to be background independent, in our sense.' The most straightforward interpretative option here, then, is to keep these two very different concepts separate: while Machian relationist theories need not be background independent according to the definitions presented in §3 (all of which cut at the distinction between dynamical versus non-dynamical structures in a physical theory), such theories *do* qualify as background independent in the sense that they do not regard as physical transformations with respect to (allegedly) unobservable structures.[144] In this sense, the Machian may also be comfortable with regarding GR as a background dependent theory.

Whatever one makes of Def. 16, it is worth also considering a more sophisticated development of this Machian proposal for background independence, suggested by Gryb:

> One important issue that the best matching framework can shed light on is that of background independence. We will show that best matching suggests a natural and precise definition of a *background*. . . . Specifically, we find it convenient to associate a background with a particular continuous symmetry of the configurations [i.e. transformations associated with the action of the similarity group]. Then, our definition implies the following: if the theory places 'physical meaning' . . . on the location of a system along a symmetry direction [i.e. along an orbit of the similarity group], then it has a background with respect to this symmetry. If it does not, then it is background independent with respect to this symmetry. Under this definition, Newtonian particle mechanics is background *dependent* with respect to rotations (it places absolute meaning to the absolute orientation of the system) while general relativity is background *independent* with respect to diffeomorphisms. (Gryb, 2010, p. 2)

Gryb's proposal for a definition of background independence can be put precisely as follows:

[143] Machians have means to address this latter issue—see below.

[144] Though, of course, it remains for the Machian relationist to make explicit the sense in which spatial distances and temporal intervals actually are unobservable—see (Pooley, 2013, §6.2).

Definition 17. (Background independence, Machian version 2) *A theory is* background independent *iff all DPMs of that theory related by the action of the similarity group are regarded as being physically equivalent.*

Def. 17 is more nuanced than Def. 16 because, on this conception of background independence, a theory need not actually be formulated on shape space in order to qualify as background independent—rather, it simply must be the case that no transformation associated with the action of the similarity group is regarded as being physically significant. Thus, on Def. 17, a theory's being formulated on shape space is a sufficient but not necessary condition for it to qualify as background independent.

Clearly, like Belot's approaches to background independence (§3.6), Def. 17 has an *interpretative* dimension: a theory is judged to be background independent insofar as DPMs lying on the same orbit of the similarity group are interpreted as being physically equivalent.[145] Though Gryb claims that such *is* the case in GR, thereby affording this theory its background independence (an important difference between Def. 17 and Def. 16), we have already seen that there are good reasons to take it to be the case that not all diffeomorphisms (that is, elements of the similarity group of GR) relate physically equivalent models—for example, we have seen in §2.2 Belot's arguments against such an interpretation of diffeomorphisms, based upon the consideration of asymptotically flat spacetimes (cf. also §5.3). Thus, even those who embrace Def. 17 need not regard GR as a background independent theory, if they resist the claim that models related by a diffeomorphism must be regarded as being physically equivalent. To put this point in the language of §2.2, while if one takes an interpretational approach to symmetry transformations GR *does* qualify as background independent on Def. 17, those who embrace Def. 17 but (arguably, reasonably) resist the interpretational approach need not arrive at this conclusion.

Finally, it is worth considering Gryb's statement that "Based on these definitions, it would seem odd even to consider [background dependent] theories as they distinguish between members of an equivalence class" (Gryb, 2010, p. 13). To unpack this statement, first recall that a symmetry transformation may (for the purposes of this piece) be regarded as an automorphism of the space of DPMs of the theory in question, such that models so related are empirically equivalent. Then consider two cases:

1. The action of the similarity group generates a symmetry transformation.
2. The action of the similarity group does not generate a symmetry transformation.

[145] Teitel (2019) argues that Belot's approach to background independence is superior, due to this interpretative dimension. Given that, he should also take seriously Gryb's approach to background independence (cf. footnote 55 of Chapter 3).

In case (1), the 'equivalence class' to which Gryb refers is the class of empirically equivalent, symmetry-related DPMs. On the interpretational approach to symmetry transformations, one need not distinguish between elements of this class—thus, Gryb's quote makes clear his tacit commitment to such a view. On the other hand, those who resist the interpretational approach need not endorse Gryb's conclusion here.

In case (2), the action of the similarity group does not generate a symmetry transformation. It is not difficult to think of theories in which such is the case—for example, in NGT, a time-dependent rotation of the matter fields can be implemented via the action of the similarity group, but models of the theory so related are not (at least given the orthodoxy) to be regarded as being empirically equivalent.[146] In this case, it is reasonable to distinguish between the worlds represented by such DPMs. Thus, in this scenario, the import of Gryb's above statement is unclear, for not only is the relevant 'equivalence class' in this case unspecified, but also, *if* this 'equivalence class' is understood to be the class of DPMs related by the action of the similarity group, then there *do* exist reasons to distinguish between elements of this class, in, e.g., NGT.

Let me summarize the findings of this section. Due to the presence of fixed structure in Machian alternatives to NGT, such theories broadly qualify as background dependent, on the definitions presented in §3. By contrast, the lack of such fixed structure means that Machian alternatives to GR broadly qualify as background independent, on those definitions. Advocates of Machian relationism propose, based upon their relative priorities in the construction of physical theories, to define background independence as, roughly, the lack of physical significance of the action of the similarity group. Though it is claimed that this approach issues the verdict that GR (even *sans* Machian reformulation) is background independent, such claims carry tacit commitment to a view akin to the interpretational approach to symmetry transformations.

4.5 Close

In this chapter, I have assessed the background independence of NGT, NCT, TPG, and KKT; my results are summarized in Table 4.1. Such work is valuable for four reasons. First, it provides a means of testing the definitions presented in §3, revealing potential shortcomings which might otherwise have gone unnoticed. For example, we saw in §4.1.5 that Def. 12 (i.e. Belot's definition of full background independence) delivers the intuitively incorrect verdict on NCT, due to issues

[146] In effect, what I am describing here is a mapping between the two models of NGT in Newton's globes thought experiment (see e.g. (Pooley, 2015, §2.2)), implemented via the action of the similarity group. For a recent article calling the orthodoxy into question, see (Gomes and Gryb, 2021).

regarding the counting of geometrical structures. In that case, I proposed a modification to the approach which resolved the issue, in the sense that the modified version of Belot's account was able to issue the intuitively correct verdict on NCT.

Second, carrying out this work allows one to move beyond mere intuitions in appraising whether the theories under consideration are background independent. For example, one might think that NCT is background independent, since it involves a dynamical spacetime as with GR. In §4.1.5, however, I found a majority verdict that NCT is not background independent, in spite of such intuitions. (By contrast, interestingly, there is a majority verdict that NCT-II is background independent.) Indeed, in cases such as KKT, it is not even clear that we *have* strong intuitions regarding background independence; the majority verdict of background independence presented in §4.3.3 is therefore useful when making judgments on this theory.

Third, this work develops our understanding of the connections between background independence and other concepts. For example, in KKT we find a theory which is (at least on some readings) not diffeomorphism invariant, yet which nevertheless arguably *is* background independent. And fourth: the analysis undertaken above leads to an enhanced understanding of the theories in question. For example, I have discussed various symmetries of NGT and NCT (§4.1.5); different classes of gauge transformation in TPG (§4.2.3); and issues regarding the diffeomorphism group associated to KKT (§4.3.3).

5
Quantum Theories of Spacetime

In this chapter, I turn my attention to quantum theories of spacetime and gravity. First, in §5.1, I think through how the definitions of background independence presented in Chapter 3, developed there in the classical content, might be modified for quantum theories. This done, I then assess the background independence of perturbative string theory (§5.2), theories associated with holographic dualities (§5.3), and loop quantum gravity (§5.4). Much ink has already been spilled regarding the background (in)dependence of such theories; my aim here is to drawn upon the work undertaken in the previous chapters in order to elevate the degree of rigor and generality of these discussions.

5.1 Quantum Background Independence

In §2.1, I presented a framework for classical theories in terms of KPMs, DPMs, and BPMs. Modifications to this framework are necessary in the case of quantum theories. In §5.1.1, I address these issues, before considering in §5.1.2 whether the definitions of background independence presented in Chapter 3 remain applicable to quantum theories. Having done this, I will be able to turn in the following sections of this chapter to the next part of the project: assessing whether various quantum mechanical theories qualify as background independent, with an ultimate view to clarifying ongoing debates regarding background independence and quantum theories of gravity.

5.1.1 Models, Reprise

The KPMs of a quantum field theory (QFT) can be taken to be tuples $\langle M, F_i, \ldots, F_m, \hat{O}_1, \ldots, \hat{O}_n \rangle$, where the F_i are fixed fields,[1] and the $\hat{O}_i$ are

[1] In the sense of **SR1**—see §3.2.2. It is necessary to include the F_i, for in e.g. standard Lorentz-covariant QFT, quantum fields are defined on a *fixed* background Minkowski spacetime.

Background Independence in Classical and Quantum Gravity. James Read, Oxford University Press.
 DOI: 10.1093/oso/9780192889119.003.0005

operator-valued fields.[2, 3, 4] Then, in such a QFT, the relevant *dynamical* question is not 'what dynamical equations do the $\hat{O}_i$ obey?', but 'what are the correlation functions of the $\hat{O}_i$?'[5] Recall that such correlations functions

$$\langle O_{\alpha_1}(x_1)\ldots O_{\alpha_m}(x_m)\rangle \equiv \langle\Omega|\, TO_{\alpha_1}(x_1)\ldots O_{\alpha_m}(x_m)\,|\Omega\rangle, \tag{5.1.1}$$

where each $O_{\alpha_i} \in \{O_1 \ldots O_n\}, i = 1\ldots m$, give the probability amplitude of finding an excitation of field O_{α_i} at spacetime location x_i $(i = 1\ldots m)$, for each of the O_{α_i}. ($|\Omega\rangle$ denotes the ground state of the (interacting) QFT in question; T is a time-ordering operator.[6]) Hence, for example, the two-point correlation function $\langle\varphi(x_1)\varphi(x_2)\rangle$ of some real scalar field φ can be interpreted (loosely) as the probability amplitude for 'propagation' of an excitation of the φ field (i.e. a φ-'particle') from the spacetime point with coordinates x_1, to the spacetime point with coordinates x_2.[7]

With the above picture of correlation functions in mind, recall now that in a QFT, such correlation functions can be computed via the *path integral*

$$\int DO_1\ldots DO_n \exp\left(i\int_{-\tau}^{\tau} d^4x\mathscr{L}\right) \tag{5.1.2}$$

—i.e. the functional integral of the exponential of a classical action (itself a spacetime integral over a Lagrangian $\mathscr{L}$)—using (see e.g. (Peskin and Schroeder, 1995, ch. 9))

[2] Those disposed toward algebraic approaches to QFT might either balk or simply be puzzled by this proposal and what's to follow. (See e.g. (Halvorson and Mueger, 2006) for an introduction to algebraic QFT.) Here I set aside discussion of debates between algebraic and 'Lagrangian-based' QFT, and proceed broadly within the framework of the latter, with its associated level of 'physics-rigor', because this will be sufficient for my purposes (cf. (Wallace, 2006)).

[3] See also the 'net assumption', discussed at (Wallace, 2006, p. 39).

[4] The introduction of F_i and $\hat{O}_i$ raises the possibility that one might also include in the definition of the KPMs of a QFT *dynamical* classical fields. Such a possibility is of relevance to §5.2 (and cf. (Rickles, 2008a)), though for simplicity it is here set aside.

[5] In what follows, hats on operators are dropped for convenience.

[6] See e.g. (Peskin and Schroeder, 1995, ch. 4). Note that one might also choose to include the quantum state in the definition of the KPMs of a QFT; I elect not to do so, as this move is unnecessary once one moves to the path integral formulation of a QFT—for which, see below. The approach that I am offering here is similar to that of Deutsch and Hayden (2000), according to which, as Wallace summarizes it, "the *state vector* is taken to be 'mere mathematical machinery': they hold it fixed throughout, writing it as $|0\rangle$ and fixing it once and for all, and independent of any contingent facts about the world, as some particular state. The operators corresponding to the dynamical variables, on the other hand, are taken perfectly seriously as the theory's ontology.... The significance of the state vector $|0\rangle$ is that it gives the probability rules: the expected value of a measurement of the field $\hat{\phi}(x,t)$, for example, is $\langle 0|\,\phi(x,t)\,|0\rangle$" (Wallace, 2012, pp. 320–321). For discussion of how the Deutsch-Hayden approach to the ontology of quantum mechanics relates to Wallace and Timpson's preferred 'spacetime state realism' (Wallace and Timpson, 2010), see (Wallace, 2012, §8.9.3).

[7] On this picture, many worlds (indeed, any consistent with the statistics specified by the correlation functions which constitute the dynamics) will be consistent with a given DPM. This is not a problem *per se*, but rather a general feature of probabilistic theories.

$$\langle O_{\alpha_1}(x_1)\ldots O_{\alpha_m}(x_m)\rangle = \lim_{\tau\to\infty(1-i\varepsilon)} \frac{\int DO_1\ldots DO_n O_{\alpha_1}(x_1)\ldots O_{\alpha_m}(x_m)\exp\left(i\int_{-\tau}^{\tau} d^4x\mathscr{L}\right)}{\int DO_1\ldots DO_n\exp\left(i\int_{-\tau}^{\tau} d^4x\mathscr{L}\right)}. \tag{5.1.3}$$

Now ask: How should DPMs of a QFT be picked out? The obvious answer is that the DPMs of a QFT are those KPMs for which correlation functions of the O_i are specified by a certain path integral. Thus, in short, DPMs of QFTs are picked out by path integrals[8]—this is the analogue of picking out DPMs via field equations, as in §2.1.[9, 10]

Finally, what of the BPMs of a classical versus a quantum theory? In this case, one can impose certain boundary conditions on the O_i of a QFT in exactly the same manner as was discussed in §2.1. Thus, no modification to the formalism of that section is necessary in this context—beyond the fact that, as mentioned, the O_i are now operator-valued fields.

5.1.2 Background Independence

The definitions of background independence presented in Chapter 3 were all constructed in the context of *classical* field theories. It is thus an open question whether they remain applicable to QFTs. Thankfully, this (broadly speaking) is the case. To illustrate, note that the definition of a fixed field (i.e. a field fixed in all KPMs of a theory—see §3.2), and of an absolute object (i.e. an object invariant across all DPMs of a theory—see §3.3), and of an absolute field (i.e. a field specified in the KPMs of a theory invariant across all DPMs of that theory—see §3.4.2) are equally applicable to quantum theories as they are to classical theories. Thus, Def. 5 (no absolute objects), Def. 6 (no formulation in terms of fixed fields) and Def. 9 (no absolute fields) can also be applied to assess the background independence of quantum theories. Indeed, it is also the case that no significant modifications need

[8] To specify a path integral, we have seen that one must specify a classical action, and so in turn a Lagrangian. However, as in classical theories, there are some QFTs which lack a Lagrangian formulation (for example $\mathcal{N} = (0, 2)$ supersymmetric QFTs in six dimensions—see e.g. (Witten, 1995; Claus et al., 1998)). Nevertheless, the framework presented in this section is capable of accommodating such theories—we simply specify that DPMs are picked out by correlation functions *directly*, without recourse to such path integrals (note that this may indeed mandate specification of the quantum state in one's KPMs). Indeed, to say that path integrals pick out DPMs in QFTs is, ultimately, a convenient shorthand for stating that DPMs are picked out by certain correlation functions.

[9] In the language of QFT, to pick out DPMs via field equations is to pick out those models which capture only the *on-shell*, or *tree-level*, contributions to the dynamics. In a QFT, we also need to capture loop-order contributions; picking out DPMs via a path integral allows us to achieve this goal.

[10] Alternatively, one could specify DPMs via the *Schwinger-Dyson equations*, which are quantum equations of motion for Green's functions derived by insisting upon stationarity of the path integral. I won't explore this idea further here, but, see (Peskin and Schroeder, 1995, §9.6) for discussion.

be made to Belot's account of background independence, beyond a reconstrual of the models of a theory *à la* §5.1.1.

The situation regarding Def. 10 (in terms of variational principles) is a little more complex. Suppose that the theory in question has an associated action. In a quantum theory, such an action can be varied to obtain the on-shell equations of motion for the objects O_i of the theory. One might worry, however, that since it is the *path integral* (rather than the action) which is fundamental in a QFT,[11] such variation of the action is irrelevant to any appraisals of the background independence of this theory.

Now, while it is true that such equations of motion do not give the *full* dynamics of the theory (unlike the classical case)—which is given by the path integral—what *is* true is that generic correlation functions in a QFT can be evaluated perturbatively,[12] with each term in such a perturbation series being constructed from *propagators*, themselves constructed from the equations of motions for the O_i of the theory which are obtained from such action principles. Thus, if a field in a QFT is subject to Hamilton's principle, this is good reason to think it dynamical in the quantum theory; if a field in a QFT is *not* subject to Hamilton's principle, then it does not appear to be dynamical in the quantum theory. Thus, the intuitions behind Def. 10 of background independence still hold in quantum theories—suggesting that it *is* still appropriate to make use of this criterion in such contexts.

It is worth mentioning one further modification to Def. 10 (also applicable to Def. 5, Def. 6, and Def. 9) which must be made when assessing the background independence of quantum theories (brought to light by de Haro et al. (de Haro, 2017a, §6.2)[13]). Since the path integral of a quantum theory decomposes into the exponential of a classical action, *and* a path integral measure, it is important that the latter object *also* contain no non-dynamical—or non-variational, etc.—fields.

Alternatively, one might pursue a more direct quantum mechanical analogue of Def. 10, posed in terms of the path integral:

Definition 18. (Background independence, path integral) *A (quantum) theory is* background independent *iff its solution space is determined by a generally covariant path integral, (i) all of whose dependent variables are integrated over in the path integral, and (ii) all of whose dependent variables represent physical fields.*

Note that Def. 18 will face some of the same potential problems as Def. 10. First: suppose that one has a Lorentz invariant QFT, featuring a Minkowski metric η_{ab} in its formulation. One could, once again following the methodology of parameterization, recast this theory in terms of clock fields, which one could

[11] Modulo the points raised in footnote 8.
[12] In terms of sums over Feynman diagrams—see e.g. (Peskin and Schroeder, 1995, ch. 4).
[13] See §5.3.3.

then integrate over in the path integral; this may once again lead to this intuitively background dependent theory qualifying as background independent—unless one can argue that the integration over those clock fields was unnecessary for the articulation of the dynamics of the theory (and one makes the modification that all dependent variables *must* be integrated over in the path integral).[14] Second: one again faces analogous issues with the construction of a path integral for theories in which boundary conditions are understood to be imposed dynamically (cf. §3.5), or for theories which simply do not admit of a Lagrangian formulation[15]—demonstrating that this definition, like its classical counterpart, can ultimately offer at best a *sufficient* condition for the background independence of a given QFT.

Many of these points apply also to Def. 15, since this definition is also cast in terms of action principles. Indeed, by analogy with our above discussion of Def. 10 versus Def. 18, I take a natural way to generalize Def. 15 to the quantum context to be the following:

Definition 19. (Background independence, path integral, corners) *A (quantum) theory is* background independent *iff (i) its solution space is determined by a generally covariant path integral, and (ii) for every $d \in Diff(M)$, the associated corner charge from the action appearing in the exponent of the path integral is non-trivial.*

Like Def. 18, Def. 19 modifies the first clause of the respective classical definition to be appropriate to path integrals (and thereby to off-shell contributions from the dynamics), while carrying over the second clause essentially intact from the classical context.

These clarifications made, I turn now to considering the background independence of three mainstream approaches to developing a quantum theory of gravity: perturbative string theory (§5.2), theories related by holographic dualities (§5.3), and loop quantum gravity (§5.4).

5.2 Perturbative String Theory

In this section, I assess whether perturbative string theory qualifies as background independent. To this end, in §5.2.1, I provide an overview of this quantum gravity

[14] Of course, all of this deserves to be worked out in detail. My point here is simply that, as Def. 18 is (I claim) the natural quantum analogue of Def. 10, it may inherit some of the problems of that classical definition.

[15] Cf. footnote 8.

theory. Then, in §5.2.2, I consider the nature of spacetime in this theory, before turning to the main task in §5.2.3. I close with some brief further remarks in §5.2.4.

5.2.1 Background

In string theory, a string can be regarded as a special case of a p-brane, which is an object with p spatial dimensions and tension $T_p = 1/(2\pi\alpha')$, where α' is the *Regge slope parameter*. The classical motion of a p-brane extremizes the $(p+1)$-dimensional volume V that it sweeps out in spacetime.[16] Thus there is a p-brane action that is given by $S_p = -T_p V$. In the case of the fundamental string, which has $p = 1$, V is the area of the string worldsheet and the action is called the *Nambu-Goto action* (Blumenhagen et al., 2013, p. 10)

$$S_{\mathrm{NG}} = -\frac{1}{2\pi\alpha'}\int_{\Sigma} d^2\sigma \left[-\mathrm{det}_{\mathfrak{ab}}\left(\frac{\partial X^a}{\partial \sigma^{\mathfrak{a}}}\frac{\partial X^b}{\partial \sigma^{\mathfrak{b}}}\eta_{ab}\right)\right]^{1/2}, \tag{5.2.1}$$

where Σ denotes the string worldsheet; the functions $X^a(\sigma, \tau)$ describe the spacetime embedding of the worldsheet; $\tau \equiv \sigma^0$ and $\sigma \equiv \sigma^1$ are coordinates on the worldsheet (the parameter τ is the worldsheet time coordinate, while σ parameterizes the string at a given worldsheet time); $d^2\sigma = d\tau d\sigma$; and Fraktur script denotes worldsheet indices. Note in (5.2.1) that the spacetime metric field η_{ab} is a fixed field in the sense of §3.2.2. Classically, the Nambu-Goto action is equivalent to the string sigma-model action (also known as the *Polyakov action*) (Blumenhagen et al., 2013, p. 13)

$$S_{\mathrm{P}(\eta)} = -\frac{1}{4\pi\alpha'}\int_{\Sigma} d^2\sigma\sqrt{-h}h^{\mathfrak{ab}}\eta_{ab}\partial_{\mathfrak{a}}X^a\partial_{\mathfrak{b}}X^b, \tag{5.2.2}$$

where $h_{\mathfrak{ab}}(\tau, \sigma)$ is an auxiliary[17] worldsheet metric with inverse $h^{\mathfrak{ab}}(\tau, \sigma)$; and $h := \det h_{\mathfrak{ab}}$. The Euler-Lagrange equation for $h_{\mathfrak{ab}}$ can be used to eliminate it from (5.2.2) and recover (5.2.1).[18] Quantum mechanically, instead of eliminating $h_{\mathfrak{ab}}$ via its classical field equations, one should perform a path integral, using standard machinery to deal with the local symmetries and gauge fixing.[19] Doing

[16] Sometimes, spacetime in string theory is referred to as *target space*. Recently, in light of so-called *dualities* in string theory (see e.g. (Rickles, 2011)), Huggett (2017) has argued that phenomenal spacetime is not equivalent to target space. I set such issues aside here, but see (Read, 2019b) for discussion.

[17] In the sense that $h_{\mathfrak{ab}}$ is a new variable, initially independent of the pullback of the spacetime metric to the worldsheet.

[18] Varying (5.2.2) with respect to $h^{\mathfrak{ab}}$, we obtain the equation of motion $\eta_{ab}\partial_{\mathfrak{a}}X^a\partial_{\mathfrak{b}}X^b - \frac{1}{2}h_{\mathfrak{ab}}h^{\mathfrak{cd}}\eta_{ab}\partial_{\mathfrak{c}}X^a\partial_{\mathfrak{d}}X^b = 0$; back-substitution then returns (5.2.1).

[19] I.e. path integral quantization *à la* Faddeev-Popov—see e.g. (Blumenhagen et al., 2013, §3.4).

this, one finds that there is a conformal anomaly[20] unless the spacetime dimension is $D=26$.[21]

For a closed string, one imposes periodicity in σ. Choosing its range to be π, one identifies both ends of the string $X^a(\tau,\sigma)=X^a(\tau,\sigma+\pi)$. After quantising and defining suitable ladder operators,[22] one can act on the ground state of the string with raising operators to study its spectrum. For the closed string, there are three distinct first excited states, denoted f_{ab}, B_{ab}, and Φ. f_{ab}—suggestively christened the *graviton*—is symmetric and traceless in its spacetime indices, and transforms under $SO(D-2)$ as a massless spin-two particle.[23] B_{ab} transforms under $SO(D-2)$ as an antisymmetric, second-rank tensor. The trace term Φ is a massless scalar, which is called the *dilaton* (Becker et al., 2007, p. 53).

In (5.2.2), I considered only a fixed, flat background η_{ab}. One can analyze more general possibilities by introducing the fields f_{ab}, B_{ab}, and Φ into the worldsheet action as background fields. The appropriate action is then (Blumenhagen et al., 2013, p. 429)

$$S_{P(\eta+f,B,\Phi)} = -\frac{1}{4\pi\alpha'}\int_\Sigma d^2\sigma\sqrt{-h}\Big(h^{\mathfrak{ab}}\partial_{\mathfrak{a}}X^a\partial_{\mathfrak{b}}X^b\left(\eta_{ab}+f_{ab}(X)\right)$$
$$+\varepsilon^{\mathfrak{ab}}\partial_{\mathfrak{a}}X^a\partial_{\mathfrak{b}}X^bB_{ab}(X)+\alpha'\Phi R(h)\Big), \tag{5.2.3}$$

where my conventions are such that $\varepsilon^{\mathfrak{ab}}=\pm1/\sqrt{-h}$. Thus, in perturbative string theory there is a move from finding the excited states of strings, to treating the associated quantum fields as *background fields* in which the dynamics of further strings can be analyzed. It is upon this move that I focus in §5.2.2.

Finally, I must discuss string interactions. In string theory, the worldsheet metric $h_{\mathfrak{ab}}$ is invariant under local changes of scale ('Weyl transformations'—already introduced in §4.4, albeit in a very different context)

$$h_{\mathfrak{ab}}(\tau,\sigma)\to\Omega^2(\tau,\sigma)\,h_{\mathfrak{ab}}(\tau,\sigma). \tag{5.2.4}$$

[20] An *anomaly* arises when a symmetry of a classical theory is not manifested in the associated quantum theory. In more technical language, though the classical action is invariant under the symmetry, the associated path integral measure—used to define the quantum theory—is not. The *conformal anomaly* arises on quantization of classical string theory, and breaks the conformal invariance of the string worldsheet. This anomaly manifests itself as an extra term in the *Virasoro algebra*, which comprises the generators of the conformal group of the string worldsheet; this extra term vanishes—thereby circumventing the anomaly—only in the case of spacetime dimension $D=26$ for the bosonic string, or $D=10$ for the fermionic string. For philosophically oriented introductions to anomalies, see (Fine and Fine, 1997) and (Huggett, 2016, §4).

[21] As mentioned in footnote 20, analogous analysis for superstrings (i.e. strings for which supersymmetry is added—either on the worldsheet as in the so-called *RNS sector*, or to the background spacetime as in the *GS sector*) gives the critical dimension $D=10$ (Becker et al., 2007, p. 7). In this chapter, I focus solely on bosonic string theory.

[22] See e.g. (Becker et al., 2007, p. 53).

[23] For how the masses of such states are determined, see e.g. (Becker et al., 2007, p. 53).

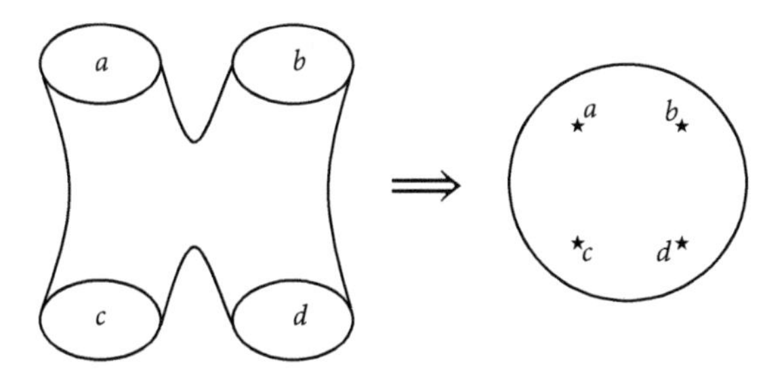

Fig. 5.1 The mapping of the four-point diagram for closed strings (left) to a compactified worldsheet (topologically a sphere) in which each of the four original outgoing world tubes is mapped to a marked point (right), via the conformal invariance of the worldsheet metric.

This invariance allows one to make conformal rescalings of the worldsheet metric; this is useful when evaluating string scattering amplitudes. For example, consider the four-point diagram for closed strings, shown on the left of Fig. 5.1.[24] By making a suitable choice of Ω, this diagram can be converted into the second diagram, in which the worldsheet is compact, the holes in the original string worldsheet corresponding to external states are closed, and the external string states now appear at marked points.[25] After rescaling, at each marked point—i.e. at each string interaction vertex—there must appear a suitable local operator with quantum numbers of the string state that was mapped to that point (Green et al., 1987, p. 34). This is a *vertex operator*, and generically takes the form

$$V_\Theta(k) = \int d^2\sigma\sqrt{-h}\,W_\Theta(\tau,\sigma)\,e^{ik\cdot X} \tag{5.2.5}$$

for emission or absorption of string state of type Θ and momentum k^a; W_Θ is a polynomial in X^a and its derivatives (Green et al., 1987, pp. 34ff.).[26] In particular, in the case where Θ is the graviton, we must pick the associated W_{grav} to saturate the two symmetric spacetime indices; the minimal such operator is $W_{\text{grav}} = h^{\mathfrak{ab}}\partial_{\mathfrak{a}}X^a\partial_{\mathfrak{b}}X^b\zeta_{ab}$, for some ζ_{ab}. Thus, we expect the vertex operator for the graviton $V_{\text{grav}}(k)$ to read

[24] Since we are now dealing with *strings* rather than point particles, we have *tubes* or *sheets* (depending on the string topology: open or closed, respectively) rather than *lines* in the analogues of Feynman diagrams.

[25] For an extended discussion of this point, see (Green et al., 1987, pp. 32ff.).

[26] The reason for the factor $e^{ik\cdot X}$ is the following. As stated by Green et al. (1987, p. 35), though the operators W_Θ transform correctly under Lorentz transformations, we must also take account of spacetime translations. Under the global symmetry $X^a \to X^a + a^a$ in which the position of each string is shifted by a constant a^a, the wavefunction of an external state of momentum k^a is multiplied by $e^{ik\cdot a}$. The simplest quantum field operator that transforms in this way under translations is $e^{ik\cdot X}$, so we postulate that this factor is present for emission or absorption of a string of momentum k^a.

$$V_{\text{grav}}(k) = \int d^2\sigma \sqrt{-h} h^{\mathfrak{ab}} \partial_{\mathfrak{a}} X^a \partial_{\mathfrak{b}} X^b \zeta_{ab} e^{ik\cdot X}. \tag{5.2.6}$$

This is understood as the vertex operator for emission or absorption of a graviton of wavefunction $\zeta_{ab} e^{ik\cdot X}$, and is discussed further in §5.2.2.

5.2.2 Fields

Background Fields and Coherent String States

One might reasonably ask: How does one bridge the conceptual gap between f_{ab}, B_{ab}, and Φ being excited states of a closed string, and these objects being *background fields*, in an action such as (5.2.3)? Heuristic attempts to answer this question have been put forward by e.g. Green et al. (1987) and Polchinski (1998); I consider here the approach of the former. Focussing on the case of augmenting (5.2.2) with an $f_{ab}(X)$ piece (i.e. ignoring for the time being contributions from B_{ab} and Φ),[27] Green et al. (1987) motivate the transition $S_{\text{P}(\eta)} \rightarrow S_{\text{P}(\eta+f)}$ by first supposing that the spacetime metric is

$$g_{ab}(X) = \eta_{ab} + f_{ab}(X), \tag{5.2.7}$$

where $f_{ab}(X)$—treated as a perturbation—represents deviation from Minkowski spacetime. The string worldsheet path integral associated to $S_{\text{P}(\eta)}$—i.e. for a string in a Minkowski background—is

$$Z_\eta = \int DX^a Dh_{\mathfrak{ab}} e^{-S_{\text{P}(\eta)}}, \tag{5.2.8}$$

while that associated to $S_{\text{P}(g)}$ is

$$\begin{aligned} Z_g &= \int DX^a Dh_{\mathfrak{ab}} e^{-S_{\text{P}(g)}} \\ &= \int DX^a Dh_{\mathfrak{ab}} e^{-S_{\text{P}(\eta)}} \left(1 + \frac{1}{2\pi} \int d^2\sigma \sqrt{h} h^{\mathfrak{ab}} \partial_{\mathfrak{a}} X^a \partial_{\mathfrak{b}} X^b f_{ab}(X) \right. \\ &\quad \left. + \frac{1}{2} \left[\frac{1}{2\pi} \int d^2\sigma \sqrt{h} h^{\mathfrak{ab}} \partial_{\mathfrak{a}} X^a \partial_{\mathfrak{b}} X^b f_{ab}(X) \right]^2 + \dots \right). \end{aligned} \tag{5.2.9}$$

The crucial step in the argument is then the following: the factor

[27] As Huggett (2016) states, dependency upon the X^a is now included so that the operator transforms correctly under translations. Cf. footnote 26.

$$V_f = \frac{1}{2\pi} \int d^2\sigma \sqrt{h} h^{\mathfrak{ab}} \partial_{\mathfrak{a}} X^a \partial_{\mathfrak{b}} X^b f_{ab}(X) \tag{5.2.10}$$

appearing in the above is the vertex operator for the emission or absorption of a graviton with wavefunction $f_{ab}(X)$![28] An insertion of V_f into the path integral (5.2.8) accommodates the interaction of strings with an external graviton of wavefunction $f_{ab}(X)$; an insertion of e^{V_f} into (5.2.8) describes the interaction of such a string with a coherent state of gravitons, and "this corresponds precisely to string propagation in the metric $g_{ab}(X) = \eta_{ab} + f_{ab}(X)$" (Green et al., 1987, p. 166)[29, 30] As Huggett states, "A coherent state is very 'unquantum', in the sense that adding and subtracting field quanta does not affect it" (Huggett, 2016, p. 7);[31] indeed, it is in this manner that *classical* background fields arise in perturbative string theory.[32]

On the basis of these results, Huggett and Vistarini conclude:[33]

> 'Background' fields do not represent new degrees of freedom in addition to those of string theory: they are not distinct primitive entities. Instead they represent the behaviour of coherent states of string excitations: the quantum states, that is, which describe classical field behaviour. (Huggett and Vistarini, 2015, p. 1169)

One must question whether this is strictly correct, for the metric field $g_{ab}(X)$, constructed as per (5.2.7), at this stage remains a *composite object*, consisting both of graviton excitations $f_{ab}(X)$ and an *ontologically primitive* background field η_{ab}. In order to vindicate claims such as the above, one must also account for the η_{ab} field. In fact, however, to do so is not straightforward. Naïvely, one might expand the $e^{-S_{P}(\eta)}$ piece in (5.2.9),[34] and attempt to consider this also as a coherent state

[28] As stated Green et al. (1987, pp. 165–166), in (5.2.6) I considered only gravitons the wavefunctions of which are plane waves, of the form $\zeta_{ab} e^{ik\cdot X}$. However, "there is no reason not to consider a wavefunction instead that is a general superposition of plane waves"—such as is the case for $f_{ab}(X)$.

[29] Notation in this quotation has been modified for consistency with this work; there is no change in content.

[30] There are questions in this vicinity regarding whether the full space of solutions of GR can be built up from a linearized theory in the manner discussed here—see, for example, the classic work of Feynman et al. (1995), and ensuing more modern discussions such as (Pitts and Schieve, 2007; Padmanabhan, 2008). Smolin (2006), argues that it is *not* the case that one can construct the full space of solutions of GR in perturbative string theory. Even if this were so, however, it is unclear (absent some further argument) why this should be construed as problematic. Indeed, in a parallel manner, Barbour has argued that it is an *advantage* of shape dynamics that it has a more restricted space of solutions than GR, for this renders the theory "more predictive" (see e.g. (Barbour, 2010), and §4.4). Though interesting, these issues shall here be set aside.

[31] See (Huggett, 2016, §2) for an extended discussion of this point.

[32] Excluding the η_{ab} field at this point—see the discussion below.

[33] See also (Huggett, 2016, §§2-3), and works by string theorists such as Horowitz (1988); Witten (1996); Seiberg (2007).

[34] With respect to the $e^{-S_{P}(f)}$ piece, i.e. run the same reasoning as above for this other term in the path integral.

of gravitons. However, this move will not work, for the η_{ab} field, unlike the $f_{ab}(X)$, is *not* trace-free.[35]

As a possible way out here, one might be tempted to argue as follows. When exploring the closed string spectrum, the first excited state is decomposed into irreducible representations of the transverse rotation group $SO(D-2)$, which are then understood to correspond to the f_{ab}, B_{ab}, and Φ fields.[36] However, there is (one might say) no *a priori* reason why this decomposition must be complete—suppose instead that the symmetric, trace-free piece f_{ab} and trace piece Φ are *not* so split, so that one considers one first excited state of the closed string to transform under $SO(D-2)$ as a massless, spin-two particle w_{ab}, *with trace*.[37] Then, the vertex operator for this object will read

$$V_w = \frac{1}{2\pi}\int d^2\sigma\sqrt{h}h^{\mathfrak{ab}}\partial_{\mathfrak{a}}X^a\partial_{\mathfrak{b}}X^b w_{ab}(X), \tag{5.2.11}$$

and therefrom a background Minkowski metric field η_{ab} *can*—it is argued—be constructed, via the above argument. Again, however, this approach faces problems, for even to *identify* the first excited states of a closed string in this manner[38] *presupposes* the existence of a Minkowski spacetime metric field η_{ab} (Blumenhagen et al., 2013, ch. 2). Thus, it does not appear that one can run an argument of this type without prior commitment to the existence of such a field. Consequently, I conclude at this stage that more needs to be done to bring Huggett and Vistarini's above claim to fruition—and that any story that one *does* tell here will be non-trivial in nature.[39]

Primitive Worldsheet Ontology

In spite of the above skepticism, it is interesting to consider what would follow *if* a string-theoretic ontological reduction *à la* Huggett and Vistarini could successfully be carried out. Supposing this to be so, it is the worldsheet metric $h_{\mathfrak{ab}}$ that

[35] *Pace* Huggett and Vistarini (2015), this is also true of a "general Lorentzian metric." Note also that speaking of such 'general Lorentzian metrics' consisting in coherent string states overlooks possible complications arising out of the discussion in footnote 30.

[36] For details, see e.g. (Blumenhagen et al., 2013, pp. 49–50).

[37] Here, however, is one reason why one might feel compelled to complete the decomposition: on the Wigner classification of elementary particles, according to which they correspond to *irreducible* representations of the groups under consideration, reducible representations are composite, and capable of further reduction into states corresponding to *fundamental* particles—here the scalar and traceless parts.

[38] And indeed in the standard manner, in terms of f_{ab}, B_{ab}, and Φ fields.

[39] One other way to proceed here would be to appeal to Brown's *dynamical approach* to fixed spacetime structure (Brown, 2005): in this context, the view would state the target space Minkowski metric field *just is* a codification of the string dynamics; it is not ontologically fundamental. Alternatively, one might wonder whether just a matrix diag $(-1,1,1,1)$ (a confined object, in Smolin's sense), rather than a full (tensorial) Minkowski metric, is required here—this kind of suggestion is made by Pitts (2012), albeit in a different context.

is fundamental, *not* the spacetime metric g_{ab}. There are two obvious attractive elements to this way of thinking:

1. On this account, *strings come first*; familiar ontology is built up from string states with no prior assumptions about spacetime of a 'non-stringy' nature.
2. The field $h_{\mathfrak{ab}}$ is invariant under (5.2.4), so a string theory can be considered a two-dimensional conformal field theory on the worldsheet. Thus, *only conformal worldsheet structure need be postulated.* This means that less metrical structure is required in string theory than might be anticipated—see §5.2.3.

Field Equations

Setting aside the above issues of ontological priority, I return now to the question of whether perturbative string theory qualifies as background independent. Here, Huggett and Vistarini write the following:

> [I]n a central sense, string theory is background independent: the metric arises from string interactions, rather than being stipulated a priori. Just like general relativity, many solutions are possible, but matter and gravity have to satisfy a mutual dynamics—except in string theory, there is no fundamental distinction between the two, a significant ontological unification.
>
> (Huggett and Vistarini, 2015, p. 1169)

To see what is meant here, consider again a string propagating in background fields, as per e.g. (5.2.3). As a worldsheet QFT, string theory is conformally invariant. It turns out—*remarkably*—that this conformal invariance *requires for consistency* that the background fields be dynamically coupled together in the Einstein equation, plus higher order corrections.[40] Thus, the 'background fields' cannot be fixed after all—they *must* be dynamically coupled via Einstein-type field equations.

5.2.3 Background Independence

In light of this result, I return now to our eight definitions of background independence—Def. 5 (no absolute objects), Def. 6 (no formulation featuring fixed fields), Def. 9 (no absolute fields), Def. 10 (in terms of variational principles), Def. 18 (in terms of the path integral), Def. 15 (in terms of non-trivial corner charges and classical action), Def. 19 (in terms of non-trivial corner charges and

[40] For original sources on this result, see e.g. (Callan et al., 1985); for textbook discussion, see (Green et al., 1987, pp. 167ff.); for presentations in the philosophical literature, see (Huggett, 2017, §3) and (Read, 2019b).

the path integral), and Def. 12 (Belot's approach)—in order to assess whether perturbative string theory qualifies as background independent, as Huggett and Vistarini claim. For completeness, I carry out these appraisals both at the level of spacetime fields and at the level of worldsheet fields.

Beginning with the former, the fields obey, as mentioned in §5.2.2, the dynamical equations of GR, plus corrections. For this reason, Def. 6, Def. 9, and Def. 12 are straightforwardly satisfied: as with GR, there are no fixed/absolute fields in the theory governing the dynamics of these fields; moreover, there appears to be a one-one match between physical and geometrical degrees of freedom.[41] That being said, it is of course also the case that Def. 5 is *not* satisfied by these background fields, for, as with GR (cf. §3.3.3), the metric determinant will qualify as an absolute object in these theories. The situation regarding Def. 10 is a little more subtle, for one must be clear on the relevant action in this case. This is *not* the worldsheet action (5.2.3); rather, the fact that spacetime fields obey the Einstein equation plus corrections means that there exists an action, variation of which yields these equations: simply the Einstein-Hilbert action, plus corrections; *this* is the relevant action in this context.[42] In such an action, all dependent variables are subject to Hamilton's principle, and there are no unphysical fields. Thus, again, it *does* appear that Def. 10 is satisfied. (Since we are dealing with *classical* equations of motion for the target space fields, Def. 18 is not readily applicable here; *mutatis mutandis* for Def. 19, although for the very reasons articulated in the context of GR by Freidel and Teh (2022), one expects that Def. 15 will be satisfied by the classical background fields here.)

Of course, all of the above proceeds on the assumption that the relevant spacetime object for consideration is g_{ab}, rather than η_{ab} and f_{ab}. If one regards the latter two objects as fundamental, then perturbative string theory, beginning with a Minkowski target space, *does* possess fixed and non-variational fields—so Def. 5, Def. 6, Def. 9, Def. 10, and Def. 15 are all violated. Indeed, if it is η_{ab} which is regarded as 'geometrical', then Belot's definition of background independence, Def. 12, is also violated, for this object is identically the same in all models of the theory. (All this is consistent with Rovelli's claim that, in this sense, perturbative string theory is not background independent: see (Rovelli, 2004, p. 9).) Note that, if *both* η_{ab} and f_{ab} are regarded as being 'geometrical' in Belot's sense, then, counter intuitively (on the assumption that f_{ab} covaries with the other, non-geometrical fields), Belot's definition will, in fact, be *satisfied*, unless one makes the same modification to this proposal as indicated in §4.1.5—i.e. unless one stipulates that *every* field designated as being 'geometrical' must covary with the matter fields; since η_{ab} does *not* do this, Def. 12 will then be violated.

[41] Modulo the potential problem cases for GR discussed in §3.6.4, which carry across to this context.

[42] Actually, I haven't proven that there invariably exists an action principle for the equations of motion for the background fields derived via the above-adumbrated string-theoretic methods. Doing so would be an interesting task; here, I simply assume that such an action exists.

Turn now to the worldsheet action (5.2.3).[43] Here, again, there are no fixed fields, absolute fields, absolute objects, or non-variational fields, for both $h_{\mathfrak{ab}}$ and the X^a are varied to yield associated equations of motion (see e.g. (Becker et al., 2007, p. 27)). Though the fields g_{ab}, B_{ab}, and Φ might *appear* to be fixed, from the worldsheet perspective these are viewed as functions of the X^a;[44] thus, they do not compromise background independence on Def. 5 (no absolute objects), Def. 6 (no formulations featuring fixed fields),[45] Def. 9 (no absolute fields), Def. 10 (variational principles), or Def. 15 (non-trivial corner charges).[46] While one might worry that the X^a fields violate the prohibition on *unphysical* fields in Def. 10, this is not so—for these functions encode precise physical information on the spacetime location of the string. In a parallel manner, while one might worry that the field $h_{\mathfrak{ab}}$ is unphysical, whether *this* is so will depend upon whether one considers $h_{\mathfrak{ab}}$ a fundamental field (§5.2.2), or an auxiliary construction (§5.2.1)—on the former view, the field is physical; on the latter, it is not. Whether one considers Def. 10 to be satisfied hinges upon this point; *mutatis mutandis* for Def. 18 (in terms of the path integral).[47]

Finally, turn to Belot's account. From the worldsheet point of view, it is natural to take the field $h_{\mathfrak{ab}}$ to constitute *geometrical structure* in the theory. But now recall again that it is possible to eliminate $h_{\mathfrak{ab}}$ from (5.2.2), to recover (5.2.1). In this case, one can envisage models of perturbative string theory which differ in the X^a, so that the spacetime embedding of the string is different, but with the *same* $h_{\mathfrak{ab}}$ field on the worldsheet—and the most natural reading of this situation is as a violation of Def. 12—*unless* one can also argue that models which differ solely in the spacetime location of the string are gauge equivalent.[48]

Thus, a detailed assessment of the background independence of perturbative string theory yields the verdict that the status of this theory is much less clear-cut than its being 'manifestly background dependent'—claims to this effect being found in e.g. (Smolin, 2008, 2006; Rovelli, 2013). The findings of this section are summarized in Table 5.1.

[43] Strictly, this is just one example of a string worldsheet action featuring background fields—however, the analysis presented here carries across more generally.

[44] Which, as we saw in §5.2.2, obey precisely specifiable relations.

[45] Again remembering that we have not demonstrated that there does not exist any reformulation of the theory which *does* involve fixed fields—which would indeed pose a problem for Def. 6.

[46] Though initially we stated that η_{ab} in (5.2.1) and (5.2.2) was considered to be a fixed field, in light of the discussion in §5.2.2, it is better to view this field as one *solution* of the more general case. Thus, this field does not compromise background independence.

[47] Since (5.2.2) is equivalent to (5.2.1), however, it may be possible to argue that even if $h_{\mathfrak{ab}}$ *is* unphysical, its presence does not compromise background independence on Def. 10—for such appraisals should be made with respect to the latter action.

[48] One might advocate this, on the grounds that the relevant issue from the worldsheet point of view appears to be whether each non-equivalent configuration of geometrical structure *on the worldsheet* is considered physically non-equivalent (recall Def. 12)—with the spacetime location of the worldsheet being irrelevant to this appraisal. If this move is defensible, then perhaps there *is* room to argue that Def. 12 is not violated, in this context.

Table 5.1 The appraisals delivered by Def. 5 (no absolute objects), Def. 6 (no formulations featuring fixed fields), Def. 9 (no absolute fields), Def. 10 (variational principles), Def. 12 (Belot's proposal), and Def. 15 (non-trivial corner charges) on the background independence of (i) perturbative string theory with respect to the target space field g_{ab} ('ST-TS-g_{ab}'), (ii) perturbative string theory with respect to the target space field η_{ab} ('ST-TS-η_{ab}'), and (iii) perturbative string theory with respect to the string worldsheet fields ('ST-W'). ✓ indicates that the theory is judged to be background independent; ✗ indicates that the theory is judged to be background dependent; and * indicates that there are some subtleties regarding the verdict.

	Def. 5	Def. 6	Def. 9	Def. 10	Def. 12	Def. 15
ST-TS-g_{ab}	✗	✓	✓	✓	✓	✓
ST-TS-η_{ab}	✗	✗	✗	✗	✓*	✗
ST-W	✗	✓	✓	✓*	✗*	✓

5.2.4 Closing Remarks

I'll close this section with two brief further remarks. First: Vistarini (2019) proposes a definition of background independence very similar to Def. 5, in terms of the absence of absolute objects (albeit without explicit reference to that definition). Although Vistarini issues an affirmative verdict on background independence in this sense, she doesn't distinguish carefully between worldsheet and spacetime readings of the theory, as we did in the previous section; she's also arguably insufficiently sensitive to the existence of absolute objects in all such formulations.[49]

Second: we saw above that the move from the excited states of a quantized string, to a coherent state of such strings in spacetime, is a rather delicate one. It is possible that the program of *string field theory*—according to which, in parallel with QFT, one envisages a field of strings in various excited states at every spacetime point—will help to flesh out this story. For now, however (and, indeed, for some decades) string field theory remains in a state of development: see (Erbin, 2021) for a recent introduction to the state of the field.

5.3 Holography

Recently, it has been claimed that background independent gravity theories can be constructed via so-called *holographic dualities*. The idea is that one begins with a Yang-Mills QFT, then translates into the language of a gravity theory via the duality. In this section, I assess whether the gravity theories so constructed are

[49] For further critical engagement with Vistarini on background independence and string theory, see (Dougherty, 2019).

indeed background independent. The structure is as follows. In §5.3.1, I introduce holographic dualities. In §5.3.2, I introduce four importantly different classes of diffeomorphism, which may be applied to certain theories related by holographic dualities. In §5.3.3, I discuss two notions of background independence found in (de Haro et al., 2016; de Haro, 2017a), which have been proposed in this context. In §5.3.4, I assess the background independence of such gravity theories: I find that one's assessment in this regard hinges upon the definition of background independence at play—though there is a majority verdict that such theories *are* background independent.

5.3.1 Holographic Dualities

The idea of holographic duality was first proposed by 't Hooft (1993). A holographic duality exists when a theory defined on a D-dimensional spacetime (the 'bulk') is theoretically equivalent[50] to a theory defined on a lower-dimensional spacetime that forms the boundary of the bulk.[51] Often, the bulk theory is a gravity theory, whereas the boundary theory is a (quantum) gauge theory.[52] In that case, a holographic duality is referred to as a *gauge/gravity duality*. Such dualities are often not only mathematically striking, but also of practical value to physicists;[53] as a result, they are studied widely in contemporary theoretical and mathematical physics.

Throughout this chapter, I deploy as a running example the best-known case of gauge/gravity duality: the *AdS/CFT correspondence* (first proposed by Maldacena (1998)), in which the bulk theory is set in anti-de Sitter (AdS) spacetime,[54] and the boundary theory is a conformal field theory (CFT). One concrete example of this correspondence is that between type IIB string theory[55] on the ten-dimensional product target space $AdS_5 \times S^5$—this being a putative quantum theory of gravity—

[50] See footnote 4 of §4.1.

[51] That the physics of the bulk is encoded in the boundary is reminiscent of holography, hence the nomenclature.

[52] Care regarding the meaning of 'gauge theory' is needed—typically, in the context of gauge/gravity duality, what is meant are Yang-Mills theories, thereby excluding other candidate 'gauge theories' such as GR. (See (Wallace, 2015) for a discussion regarding the parallels between these two types of theory; cf. also (Teh, 2016).)

[53] For example, it is often the case that the strong coupling regime of one theory—in which problems are difficult to solve by perturbative methods—can be translated into the weak coupling regime of its dual, in which problems are more tractable. See e.g. (de Haro et al., 2016, p. 1382).

[54] By this, I mean a Lorentzian manifold $\langle M, g_{ab}\rangle$, the metric field g_{ab} for which is a solution to the vacuum Einstein equation with negative cosmological constant, $G_{ab} + \Lambda g_{ab} = 0$ ($\Lambda < 0$). One also needs to impose boundary conditions: this will be discussed below.

[55] One of the five consistent superstring theories—see e.g. (Becker et al., 2007, pp. 8ff.), and footnote 21 of §5.2.1.

and $\mathcal{N} = 4$ supersymmetric Yang-Mills theory[56] on the four-dimensional boundary of the AdS_5 space—this being a supersymmetric CFT.[57]

Let me give further detail on bulk theories in the AdS/CFT correspondence. Importantly, the low-energy approximation to the bulk equations of motion is given by Einstein's equation with negative cosmological constant, $G_{ab} + \Lambda g_{ab} = 8\pi T_{ab}$ $(\Lambda < 0)$ (de Haro et al., 2016, p. 1394). For any space which satisfies these equations, and given a conformal metric at infinity,[58] the associated line element can be written in the *Fefferman-Graham form*,[59] (de Haro et al., 2016, p. 1394)

$$ds^2 = \frac{l^2}{r^2}\left(dr^2 + g_{ij}(r,x)\,dx^i dx^j\right), \tag{5.3.1}$$

where $g_{ij}(r,x)$ is an arbitrary function of the *radial coordinate*, r. The remaining coordinates x^i $(i = 1,\ldots,d := D-1)$ parameterize the boundary, which is of dimension d, and located at $r \to 0$.[60] The conformal metric at the boundary is $g_{(0)ij}(x) := g(0,x)$.[61] Solving Einstein's equations amounts to finding $g_{ij}(r,x)$ given some initial data; because of the presence of a timelike boundary, choosing a spatial surface at some initial time does not completely specify the problem (de Haro, 2017a, §2).[62] In addition, we must provide boundary conditions. $g_{ij}(r,x)$ has a regular expansion in a neighborhood of $r = 0$ as

$$g_{ij}(r,x) = g_{(0)ij}(x) + r g_{(1)ij} + r^2 g_{(2)ij} + \ldots. \tag{5.3.2}$$

The coefficients in this expansion, apart from $g_{(0)ij}$ and $g_{(d)ij}$, are all determined algebraically from Einstein's equations. By contrast, the coefficients $g_{(0)ij}$ and $g_{(d)ij}$ are not so determined[63]—these are the boundary data.[64]

[56] $\mathcal{N}$ denotes the number of supersymmetry generators in a QFT.

[57] As de Haro (2017a) states, generically, neither is the bulk pure $AdS_5 \times S^5$ nor is the boundary theory exactly conformal—it is sufficient that the bulk be asymptotically locally $AdS_5 \times S^5$ and that the QFT on the boundary have a fixed point. (Recall that in QFT, renormalization group flow may take us to fixed points, which are CFTs.) This is discussed further below. Note also that, in the string-theoretic context, we are not considering here the effective spacetime obtained via the beta function analysis presented in §5.2, but rather the 'fixed' background on which the string theory is defined initially.

[58] This being necessary for the formulation of the AdS/CFT correspondence. By 'conformal metric at infinity', I mean that the boundary metric is a representative of a conformal equivalence class of metrics (Ammon and Erdmenger, 2015, p. 209).

[59] Computed originally by Fefferman and Graham (1985).

[60] See (de Haro et al., 2016, §2).

[61] Cf. footnote 58.

[62] I do not say 'Cauchy surface', since AdS space is not globally hyperbolic.

[63] Only the trace and divergence of $g_{(d)ij}$ is determined from Einstein's equations (de Haro et al., 2016, p. 1395).

[64] We recover pure AdS space when $g_{(0)ij}(x)$ is chosen to be flat (cf. footnote 57). Though $g_{(0)ij}$ and $g_{(d)ij}$ are at the outset arbitrary and unrelated, there do exist relations between the two quantities—see (de Haro et al., 2016, p. 1395). For original sources, see (de Haro et al., 2001, §2).

5.3.2 Bulk Diffeomorphisms

Before proceeding further in this discussion of the background independence of theories related by holographic dualities, I must distinguish four distinct classes of diffeomorphism, applied to the bulk theories under consideration. These are:[65]

- **(a1-1A)** Diffeomorphisms ϕ which preserve the asymptotic form of the metric field $g_{(0)ij}$, and which are asymptotic to the identity at infinity—that is, $\phi|_{\text{boundary}} = \text{Id}$.[66]
- **(a1-1B)** Diffeomorphisms ϕ which preserve the asymptotic form of the metric field (so that $\phi^* \left(g_{(0)ij}\right) = g_{(0)ij}$), but which are not necessarily asymptotic to the identity at infinity (so that in general $\phi|_{\text{boundary}} \neq \text{Id}$).[67]
- **(a1)-2** Diffeomorphisms ϕ which do not preserve the asymptotic form of the metric field (so that $\phi^* \left(g_{(0)ij}\right) \neq g_{(0)ij}$), but which do keep the boundary metric in its conformal class, so that $\phi^* \left(g_{(0)ij}\right) \in \left[g_{(0)ij}\right]$, where $\left[g_{(0)ij}\right]$ denotes the conformal equivalence class of boundary metric fields associated with $g_{(0)ij}$.[68]
- **(a2)** Diffeomorphisms ϕ which change the conformal class of the asymptotic metric field, so that $\phi^* \left(g_{(0)ij}\right) \notin \left[g_{(0)ij}\right]$.

Both diffeomorphisms of type (a1)-1A and (a1)-1B preserve the asymptotic form of the metric field; while the former act trivially on the boundary metric, the latter in general do not. Diffeomorphisms of type (a1)-2 change the asymptotic form of the metric, but keep it within its conformal class. Diffeomorphisms of type (a2) change the conformal class of the asymptotic form of the metric field.

For Belot, as we have already seen in §§2.2, 3.6.3, two DPMs of GR with asymptotic boundary conditions are gauge equivalent if and only if they are related by a diffeomorphism asymptotic to the identity at spatial infinity.[69] Belot (2018) lists a number of advantages to this approach (see e.g. the quotation given in §2.2), chief among which is that not regarding diffeomorphisms of type (a1)-1B as relating physically equivalent models allows for the construction of conserved quantities in the theories under consideration:[70]

[65] This notation is an extension of that introduced in (de Haro, 2017a), and is chosen for consistency with that work.

[66] For Belot, these are diffeomorphisms of type D_0—cf. §2.2 and (Belot, 2008, p. 201).

[67] For Belot, these are diffeomorphisms of type $D = D_0 \times G$—cf. §2.2 and (Belot, 2008, p. 201).

[68] Clearly, these diffeomorphisms implement conformal transformations of the boundary metric.

[69] Belot's discussion is also applicable to the bulk theories related by holographic dualities under consideration in this chapter.

[70] On this issue, Belot quotes with approval similar comments made by Penrose (1982, p. 632).

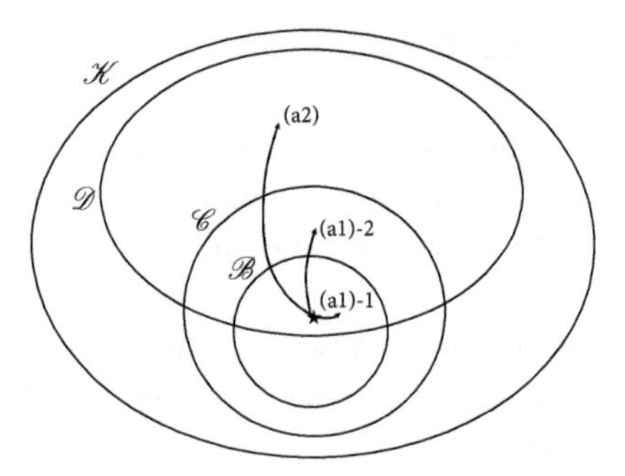

Fig. 5.2 The action of diffeomorphisms of types (a1)-1 (both (a1)-1A and (a1)-1B), (a1)-2, and (a2) in the space of models of a bulk theory in an AdS/CFT duality.

> In the regime in which asymptotic flatness is imposed at spatial infinity, why think of the true degrees of freedom as being given by the space of solutions modulo the relation of being related by an asymptotically trivial [diffeomorphism], rather than [modulo] the relation of being related by any old [diffeomorphism]? One powerful consideration is that the Poincaré group... acts on the former space in an interesting way (its action on the latter space is trivial)—and that one finds in this case the usual rich and fruitful relation between symmetries of a physical theory and conserved quantities. Thus, corresponding to any family of time translations at infinity will be a notion of the energy of an instantaneous state, which will be conserved over time in any asymptotically flat solution. With this notion in hand, we can, e.g. note that the physics of isolated systems is time-translation invariant in general relativity (with vanishing cosmological constant)—for any solution, you can find another differing from it by a time translation at infinity, in which the same instantaneous states occur in the same order. (Belot, 2018, pp. 967–968)

Note that this position is *stronger* than that of de Haro, who argues that only select diffeomorphism of type (a1)-2 (cf. §5.3.4), and diffeomorphisms of type (a2), should be regarded as relating physically inequivalent models.[71] For de Haro, such diffeomorphisms should be regarded as relating physically inequivalent models, as salient physical quantities (notwithstanding conserved quantities, as discussed by Belot) differ between those models—see e.g. (de Haro, 2017a, §2.3.2).

Let us attempt to visualize these different types of diffeomorphism. As stated by de Haro, "in order to obtain a solution of the equations of motion of general

[71] Strictly, de Haro is silent (in e.g. (de Haro, 2017a; de Haro et al., 2017)) on whether diffeomorphisms of type (a1)-1B should be regarded as relating physically inequivalent models. However, in the remainder of this chapter, I understand him to endorse the position that only diffeomorphisms of type (a1)-2 (in some circumstances—see below) and (a2) be regarded as relating physically inequivalent models.

relativity in spaces with a negative cosmological constant, boundary conditions need to be specified" (de Haro, 2016, p. 2). Hence, for the case in which the theory in question is GR with negative cosmological constant, to pick out any point in $\mathscr{D}$, one must specify boundary conditions. Now, choose one such point in $\mathscr{D}$. Then, call the class of models which obey the same boundary conditions $\mathscr{B}$, and the class of models which have the same conformal class of boundary conditions $\mathscr{C}$. The picture that emerges is as per Fig. 5.2.[72] On this setup, transformations of type (a1)-1 (both (a1)-1A and (a1)-1B) keep us in $\mathscr{B}$; generic transformations of type (a1)-2 take us to a point out of $\mathscr{B}$ but still in $\mathscr{C}$; and transformations of type (a2) take us to a point out of $\mathscr{B}$ to a point in $\mathscr{D}$ and not in $\mathscr{C}$.

5.3.3 Background Independence

It is sometimes claimed (see e.g. (Maldacena, 2005, p. 61)) that a holographic duality can be used to *define* a quantum gravity theory via the associated boundary theory.[73] Assuming this can be done, one must ask whether this quantum gravity theory manifests all (putatively) necessary/desirable qualities of such a theory, potentially including background independence. This question—are holographic bulk theories background independent?—I address in §5.3.4. Before doing so, however, I discuss two definitions of background independence due to de Haro et al., developed in the context of gauge/gravity dualities (de Haro et al., 2016; de Haro, 2017a,b).

Minimalist Background Independence

De Haro et al. define *background independence in the minimalist sense* as follows: (de Haro et al., 2016, §6.2)

Definition 20. (Background independence, minimalist): *A theory is* background independence in the minimalist sense *if:*

(a) There is a generally covariant formulation of the dynamical laws of the theory that does not refer to any background metric (or background fields). 'Dynamical laws' is here understood in terms of an action and a corresponding path integral measure; and 'background' refers to fields whose values are not determined by corresponding equations of motion.

(b) The states and the quantities are invariant (or covariant, where appropriate) under diffeomorphisms, and also do not refer to any background metric.

[72] Here, by extension from Fig. 2.1, I draw $\mathscr{B}$ and $\mathscr{C}$ overlapping the border of $\mathscr{D}$, to allow for the possibility that the fields of the theory may obey some other dynamics yet nevertheless satisfy the same boundary conditions—though this possibility will not be relevant in what follows.

[73] It is sometimes further claimed that the bulk theory *emerges* from the boundary theory—a notion subject to criticism in e.g. (Teh, 2013; Dieks et al., 2015).

Let me comment on (a) and (b). Beginning with the former, de Haro et al. are explicit that this criterion is to be understood as an extension of Def. 10 (i.e. Pooley's definition in terms of variational principles) to quantum theories (see (de Haro et al., 2016, p. 1399); (de Haro, 2016, p. 3))—in the same spirit as Def. 18 (my proposed definition in terms of path integrals). This becomes clear, once one recognizes (I) the requirement of the existence of a generally covariant action for the theory in question; (II) the prohibition of 'background fields'—which are intended to be understood as those fields which violate (i) and (ii) in Def. 10; and (since we are now dealing with quantum theories—see §5.1.2) (III) the inclusion of a requirement that the path integral measure not make reference to any background fields.

One worry does, however, exist regarding the overlap between 'background fields' in the sense of (a) in Def. 20, and fields which violate (i) and (ii) in Def. 10. As stated above, 'background field' refers only to "fields whose values are not determined by corresponding equations of motion"—and this coincides only with (i). Values of fields such as Θ^{abcd} in the case of **SR2** *are* "determined by corresponding equations of motion," so this field does *not* qualify as a 'background field', meaning that (a) *is* satisfied in the case of **SR2**. As we saw in §3.5, this is intuitively the incorrect result. This suggests that (a) in Def. 20 should be fixed to include a prohibition on 'unphysical fields', *à la* (ii) in Def. 10. Once such an amendment is made, (a) in Def. 20 *can* be understood as a reformulation of Def. 10 in the context of quantum theories.

Now consider (b); as stated by de Haro et al. (2016), this condition is "novel." The first thing to note here is that "states and quantities" is understood to mean *physical* states and quantities—i.e., in the language of §2.1, those encoded exclusively in the elements of $\mathscr{K}$ of the theory in question.[74] The second thing to note about (b) is that it carries a tacit restriction to *relevant* diffeomorphisms (de Haro, 2017a, p. 114)—by which are meant those diffeomorphisms understood to be gauge transformations. With this in mind, (b) can be read as follows: 'Physical states and quantities must be invariant under those diffeomorphisms that are understood to be gauge transformations.'[75]

Having been re-posed thus, one might ask of (b): in what sense is this condition relevant to the notion of *background* in a physical theory? To answer this question, one must recognize that, for de Haro et al., minimalist background independence is "a *necessary* requirement on any theory of quantum gravity which wishes to reproduce general relativity in the semi-classical limit" (de Haro, 2016, p. 3). Part

[74] For a more precise definition of 'states and quantities', see (de Haro et al., 2017, §3.2).

[75] Why the further qualification in (b) of 'or covariant, where appropriate'? Such a qualification is introduced in light of issues pertaining to the *diffeomorphism anomaly*—see below. However, in my view such a qualification is unnecessary (and, arguably, involves a conflation of physical quantities, and their representation in a coordinate basis)—the key criterion of (b) is that physical quantities do not change between gauge-equivalent, diffeomorphism-related models.

of such a requirement is that physical quantities be invariant under those diffeomorphisms understood to be gauge transformations—thus justifying the inclusion of (b). Since, however, the notion of background independence discussed in this book regards the presence of 'background structure' in a theory, rather than compatibility with GR in the above sense, (b) can largely be set aside in the present analysis.[76]

Extended Background Independence

In addition to Def. 20, de Haro et al. introduce *extended background independence:*[77]

> In the extended sense of [background independence], the initial or boundary conditions are also to be dynamically determined. This would be the case in a theory with either no boundary conditions at all, or boundary conditions which are determined by the dynamical equations. (de Haro, 2016, p. 4)

Thus, extended background independence captures the desideratum that "boundary conditions should now be obtained dynamically rather than by stipulation" (de Haro, 2017a, p. 115). Extended background independence is manifestly in line with Smolin's requirement that all 'background structures' ultimately be excised from our physics (Smolin, 2003, p. 204).

5.3.4 Background Independence and Holography

Having presented these definitions of background independence due to de Haro et al., I consider now whether bulk theories in holographic dualities satisfy both these definitions, and the other definitions of background independence presented up to this point.

Minimalist Background Independence

Do bulk theories in AdS/CFT dualities satisfy Def. 20? At the classical level, we have seen in §5.3.1 that the dynamical equations of these theories are Einstein's equation with negative cosmological constant; this can be derived from a generally covariant action. Moreover, as de Haro states, "adding (string-theoretical) classical matter fields does not change this picture: since these contribute covariant terms to the equations of motion" (de Haro, 2017a, p. 113). Indeed, even adding quantum corrections does not alter this verdict, for "In the full quantum version of

[76] Though, for completeness, I *do* assess whether (b) is satisfied by bulk theories in holographic dualities. Note also that this is not to question the value of the program of de Haro et al.

[77] See also (de Haro, 2017a, §2).

AdS/CFT, quantum corrections manifest themselves as higher-order corrections in powers of the curvature to the classical action. These are generally covariant as well." Since none of these correction terms involve fields which violate (i) and (ii) in Def. 10, one might conclude that such bulk theories *do* satisfy (a) of Def. 20.[78]

This being said, one must recall here subtleties regarding theories with boundary conditions, discussed in §3.5 in the context of Def. 10. First, let the KPMs of the bulk theories (for simplicity, considered here at the classical level) be the KPMs of vacuum GR, with negative cosmological constant. Let the DPMs be those KPMs which satisfy Einstein's equation; and let the choice of boundary conditions—fixed by $g_{(0)ij}$ and $g_{(d)ij}$—fix a $\mathscr{B} \subset \mathscr{K}$.

At this juncture we face three interpretive options, depending upon how a *theory* is defined. First, one might state that a theory plus boundary conditions is not itself a new theory; background independence of a theory plus boundary conditions should be assessed with respect to $\mathscr{D} \subset \mathscr{K}$. Second, one may consider the $\mathscr{B}$ picked out by the boundary conditions to specify the KPMs of a new theory. Then, background independence should be assessed with respect to whether the dynamical laws which pick out $\mathscr{D} \cap \mathscr{B} \subset \mathscr{B}$ can be obtained as per Def. 10. Third, one may impose boundary conditions at the level of dynamics; background independence should then be assessed with respect to whether the dynamical laws which pick out $\mathscr{D} \cap \mathscr{B} \subset \mathscr{K}$ can be obtained as per Def. 10.[79]

In the first of these cases, since GR is background independent on Defs. 10 and 18, it is clear that the bulk theories in question are also background independent in this sense, and so satisfy (a) of Def. 20. As we saw in §3.5, in the second case, the definition of background independence is *also* satisfied by these theories with boundary conditions, for the very same action principle suffices to pick out $\mathscr{D} \cap \mathscr{B}$, when KPMs are restricted to $\mathscr{B}$.[80] On the third conception, however, it is *prima facie* unclear whether such theories will satisfy Def. 10, and so (a) of Def. 20, for it is unclear whether there exists an action principle or path integral which picks out $\mathscr{D} \cap \mathscr{B} \subset \mathscr{K}$. While de Haro argues that, since boundary conditions must be imposed[81] in *every* element of $\mathscr{D}$ for the bulk theories in question, it is not correct to regard each choice of boundary conditions as picking out a different theory (thereby favoring the first interpretive route above) (de Haro, 2016, pp. 2, 9), I wish to highlight here that there does exist room in logical space to make an alternative move, on which Defs. 10 and 18, and so (a) of Def. 20, may be violated.

[78] Note here the further assumption that these corrections do not lead to violations of Def. 20 at the level of the path integral measure.

[79] See §3.5.

[80] Although the *theory* in this case is not diffeomorphism invariant—for diffeomorphisms which do not preserve the boundary conditions will take one out of $\mathscr{B}$, *a fortiori* out of $\mathscr{K}$—the relevant point from the point of view of satisfaction (or otherwise) of Def. 20 is that the action principle of this theory *is* still diffeomorphism invariant.

[81] I.e. $g_{(0)ij}$ and $g_{(d)ij}$ must be specified.

Turn now to the four classes of diffeomorphism presented in §5.3.2. While de Haro adopts a standard view of diffeomorphisms of type (a1)-1 (both (a1)-1A and (a1)-1B) as relating gauge-equivalent models,[82] he also endorses a line according to which diffeomorphisms of type (a2) relate physically *inequivalent* models (de Haro, 2017a, p. 114). The situation regarding diffeomorphisms of type (a1)-2 is more subtle. As discussed at (de Haro, 2017a, §2.3.3), for odd *d*, bulk theories manifest asymptotic conformal invariance; as a result, physical quantities are invariant under transformations of type (a1)-2. For even *d*, however, "the asymptotic conformal invariance is broken by the regularisation of [a] large-volume divergence.... As a consequence, physical quantities such as the stress-energy tensor... no longer transform covariantly, but pick up an anomalous term" (de Haro, 2017a, p. 115). This is the *diffeomorphism anomaly*.

What should one make of these results? First, if, in line with de Haro, one regards diffeomorphisms of type (a2) as not being gauge transformations, then it is clear that these transformations are not *relevant* in the sense of §5.3.4. Therefore, the fact that physical states and quantities are not preserved under such transformations does not lead to a violation of (b). Similarly, if diffeomorphisms of type (a1)-2 in the case of even *d* are not regarded as gauge transformations, then they are also excluded from consideration in assessments of whether (b) in Def. 20 is satisfied—so the fact that physical states and quantities are not preserved under such transformations again presents no violation of this condition. This, indeed, is the preferred move of de Haro et al. (de Haro, 2016, 2017a; de Haro et al., 2016, 2017).[83, 84] Thus, one *can* tell a plausible story regarding the satisfaction of Def. 20 by bulk theories in AdS/CFT dualities.[85]

Finally, it is worth commenting on whether bulk theories in AdS/CFT dualities satisfy extended background independence, as presented in §5.3.3. De Haro points out that the straightforward answer here, as with GR, is *no*, for boundary conditions in such theories are *not* dynamically determined (de Haro, 2017a, §2.3.4). As de Haro argues, however, this result should not be construed as problematic, for extended background independence is best considered an heuristic *desideratum* on a quantum gravity theory, rather than as a necessary condition (de Haro, 2017a, §2.3.4).[86]

[82] *Pace* Belot—see §2.2.

[83] Note that such a move is not possible on the interpretational approach to symmetries, presented in §2.1.

[84] One might (legitimately) worry, however, that making this move *post facto* is too easy, and that some prior demarcation of which transformations qualify as physical is required.

[85] Though, of course, if one *does* maintain that diffeomorphisms of type (a1)-2 in the case of even *d* are gauge transformations, then (b) in Def. 20 is violated.

[86] In this book, I leave open whether *any* account of background independence is strictly a necessary condition on a quantum gravity theory.

Other Approaches

Above, we have seen that there is a plausible route to arguing that Def. 20 is satisfied by bulk theories in AdS/CFT dualities—and since, as stated in §5.3.3, (a) in Def. 20 is understood as a reformulation of Def. 10 in the sense of §5.1.2, Def. 10 is also satisfied by these theories—as is Def. 18. This being the case, turn now to our familiar Def. 5 (no absolute objects), Def. 6 (no formulations featuring fixed fields), Def. 9 (no absolute fields), Def. 12 (Belot's proposal), and Def. 15 and Def. 19 (both in terms of the presence of non-trivial corner charges). On Def. 5, Def. 6, and Def. 9, it is clear that evaluations of bulk theories with respect to these definitions must be made on a case-by-case basis. Nevertheless, none of the best-known bulk theories in AdS/CFT dualities—understood as classical theories plus quantum corrections—involve fixed or absolute fields;[87] this is good *prima facie* evidence that Def. 6 and Def. 9, respectively, are satisfied by these theories.[88]

On Def. 5, the bulk theories under consideration do not satisfy this definition, for these theories can be understood as vacuum GR with negative cosmological constant, plus corrections. However, recall from §3.3.3 that GR itself was found to have an absolute object—the metric determinant; given this, one naturally also expects these bulk theories to manifest an absolute object. This, indeed, is exactly the same point that was made in the context of perturbative string theory, in §5.2.3.

Consider now Belot's account of background independence. To establish which of the definitions in §3.6.2 is satisfied by bulk theories in AdS/CFT dualities, first recall that Belot understands boundary conditions to be part of the *definition* of a theory;[89] this permits him to disregard diffeomorphisms which do not preserve specified boundary conditions (e.g. those of type (a1)-2 and (a2)). Then, Belot states that diffeomorphisms of type (a1)-1B do *not* relate physically equivalent models. On this interpretation of bulk theories in holographic dualities, such theories should (as in the case of GR with asymptotic boundary conditions) qualify as *nearly background independent*, for the quotient of the group of all diffeomorphisms by the group of diffeomorphisms which act trivially on the boundary conditions is finite-dimensional. On the other hand, if one follows de Haro in regarding *all* diffeomorphisms of type (a1)-1 (both types (a1)-1A and (a1)-1B) as relating physically equivalent models, then one will—by analogy with the discussion in §3.6.3—consider bulk theories to be *fully* background independent.

Finally, what of the background independence of bulk theories in holographic dualities according to the corner charge proposal of Freidel and Teh—i.e. Def. 15? Suppose we are working in asymptotically AdS space, as discussed above. Then, there are transparently asymptotic symmetries of the metric (and, let us assume,

[87] See e.g. (Ammon and Erdmenger, 2015).

[88] Moreover, since these theories can be understood as GR plus corrections, there is also reason to suppose that alternative formulations of these theories featuring fixed fields do not exist—cf. §3.4.1.

[89] Accordingly, Belot can be understood as following either the second or third interpretative branch above.

of the whole theory)—e.g. of type (a1-1B). Insofar as we stipulate that we are working within such spacetimes, we are thereby insisting that the metric field have certain asymptotic isometries. But in that case, a similar analysis to that presented in §4.3.1 follows: associated with the diffeomorphisms which generate these isometries will be corner charges which need not vanish, but which are non-dynamical, and therefore 'trivial', by the standards of Def. 15. Thereby, it appears that the *ab initio* restriction to such spacetimes at the level of KPMs spoils the background independence of such theories, according to Def. 15.[90]

5.3.5 Coda

Having assessed the background independence of bulk theories in AdS/CFT dualities, I close by considering some further issues related to the above discussion. First, I return to the issue of the definition of general covariance (§3.1), in light of comments made by de Haro (2017a). Next, I assess whether one can exploit holographic dualities to render background independent theories which are *prima facie* background dependent.

General Covariance

De Haro (2017a) claims that the presence of the diffeomorphism anomaly "contradicts the claim, often seen in discussions of background independence (and going back to Kretschmann's objection to Einstein's claims that general covariance selected general relativity), that 'any' theory can be given a manifestly covariant formulation" (see also (de Haro et al., 2016, p. 1421)). This is questionable, as it appears to conflate the trivial notion of general covariance discussed in §3.1, with more substantive notions, such as diffeomorphism invariance (see §3.2).

Since a bulk theory typically will be formulable in the coordinate-free language of differential geometry (notwithstanding the diffeomorphism anomaly), such a theory *can* be made generally covariant in the trivial sense. On the other hand, it is true that the diffeomorphism anomaly breaks the diffeomorphism invariance of the theory. It is wrong, however, to conflate these two senses of 'general covariance', and thereby to conclude that the diffeomorphism anomaly provides a counterexample to the Kretschmann point.[91]

Inter-Theory Translation and Background Independence

As a second point of interest, I reflect on the possibility of inter-theory translation between the CFT and the bulk theory in an AdS/CFT duality. Suppose we accept

[90] For more on asymptotic symmetries and corner charges, see (Ciambelli, 2022).

[91] For related discussion regarding conflating these two different senses of general covariance, see (Pitts and Schieve, 2007).

that the bulk theory in such a duality is background independent. As de Haro states, the CFT, by contrast, violates Def. 20 of background independence, not least because in this theory "the metric field is not determined by the equations of motion" (de Haro, 2017a, p. 113)—thereby violating (a).

Given this, we appear to have a duality according to which one of the theories is background independent, and the other is not. In itself, this is an interesting result. One might, however, go further, by claiming that the CFT can be *rendered* background independent via holography (see e.g. Rozali (2009)).[92] Here, however, I urge caution, for the same reasons as given in preceding sections of this book: the two dual theories are *prima facie* very different, and (absent an overarching theory in which degrees of freedom of both can be defined, or some account as to how just *one* of the theories is metaphysically privileged—see e.g. (Read, 2016; Le Bihan and Read, 2018)) should be evaluated *on their own terms*.[93] If one follows this mantra, then one *cannot* straightforwardly regard the CFT as being rendered background independent via the holographic duality.

5.4 Loop Quantum Gravity

In this section, I appraise the background independence of *loop quantum gravity* (LQG)—a quantum theory of gravity in which the Lorentzian manifolds $\langle M, g_{ab} \rangle$ of GR (g_{ab} obeying the Einstein equation (2.1.1)) are replaced with discrete, graph-theoretic structures, obeying quantum mechanical dynamics (that is, in terms of the degrees of freedom of which quantum mechanical transition amplitudes can be computed—recall §5.1). In a straightforward sense, LQG is a quantization of GR, and thereby qualifies as a quantum theory of gravity; claims that LQG is a leading such candidate theory are, its proponents maintain, vindicated in e.g. the applicability of the theory to cosmological scenarios (Ashtekar et al., 2008; Ashtekar, 2009a,b; Rovelli and Vidotto, 2015), and its (allegedly) having GR as a classical limit (Rovelli and Vidotto, 2015, ch. 8).[94]

Much is often made of the background independence of LQG—see e.g. (Rovelli, 2001; Smolin, 2006, 2008); in particular, it is regularly claimed that LQG is *more* background independent than perturbative string theory (cf. §5.2). As an

[92] In a sense, this is the reverse of the claim often found in the literature that a quantum gravity theory—here, the bulk theory—*emerges* from the CFT, via the holographic duality. For critical discussion, see e.g. (Teh, 2013).

[93] C.f. §§3.8, 4.2.3, 4.3.3.

[94] In fact, understanding the classical limit of LQG is a subtle business—see (Crowther, 2016; Wüthrich, 2006). One reason for this is that in this case "there does not appear to be a notion of emergence that might be based on the idea of a limiting relation in the theory (or, at least, not based on a limiting relation alone)" (Crowther, 2016, p. 194). For some discussion of the classical limit of LQG from the point of view of coherent states (cf. §5.2), see (Rovelli and Vidotto, 2015, ch. 8); in this chapter, I largely set aside these issues, and, rather, evaluate LQG on its own terms.

illustration of such claims, consider the following remark, made by Matsubara: "Still, it is fair to say that string theory would be background independent in a weaker and less explicit sense compared to LQG, which is manifestly background independent" (Matsubara, 2017, p. 3). Two question immediately come to mind on being confronted with this passage:

1. By what standards is LQG assessed to be "manifestly background independent"?
2. Is it truly the case that LQG is *more* background independent than (e.g.) perturbative string theory?

The use of 'manifest' in (1) is perplexing, since authors writing on this topic in the context of LQG often do not make precise the notion of background independence under consideration. Part of the purpose of this section is to fill that lacuna, by applying to LQG the above-constructed definitions of background independence. Having done so, I will be in a better position to address (2)—i.e. to compare the background independence of LQG and perturbative string theory.

The plan for the section is as follows. In §5.4.1, I recall the essential details of LQG. In §5.4.2, I assess the background independence of this quantum theory of gravity. In §5.4.3, I close with some comparative discussion.

5.4.1 Background

Here, I introduce the essential background details of LQG—in particular, the Hilbert space of the theory, spin network states in this Hilbert space and spinfoams, and the dynamics of the theory.

Hilbert Space

The Hilbert space $\mathscr{H}$ on which LQG is built is the direct sum of 'graph spaces' $\mathscr{H}_\Gamma$,

$$\tilde{\mathscr{H}} = \bigoplus_{\Gamma} \mathscr{H}_\Gamma \qquad (5.4.1)$$

factored by an equivalence relation, '$\sim$', to be introduced shortly (Rovelli, 2010, p. 1). Thus, schematically, we have $\mathscr{H} = \tilde{\mathscr{H}} / \sim$. The sum in (5.4.1) runs over *abstract graphs*, Γ. Such a graph is defined as a set L of *links*, and a set N of *nodes*, together with two functions, assigning a source node $s(l) \in N$ and a target node $t(l) \in N$ to every link $l \in L$. Defining

$$G := \frac{SU(2)^L}{SU(2)^N}, \qquad (5.4.2)$$

the graph Hilbert space $\mathscr{H}_\Gamma$ is in turn defined to be[95]

$$\mathscr{H}_\Gamma := \langle L^2(G,\sigma,\mu), +, \cdot, \langle\cdot|\cdot\rangle\rangle. \tag{5.4.3}$$

Here, $L^2(G,\sigma,\mu)$ denotes the square-integrable functions on G, σ denotes a σ-algebra on G, and μ is the so-called *Haar measure* defined in terms of σ, via which are constructed the Lebesgue integrals in terms of which the square-integrability of functions on G is established.[96] $+$ and $\cdot$ denote (respectively) addition and multiplication operations defined on $L^2(G,\sigma,\mu)$, and $\langle\cdot|\cdot\rangle$ is an inner product on $L^2(G,\sigma,\mu)$, again defined using μ.[97] The Haar measure is the unique measure on G which is invariant under so-called *left translations* and *right translations*.[98] The details are inessential for our purposes; suffice it to note that this invariance affords the Haar measure a number of useful properties in the study of LQG.

$\mathscr{H}$ is obtained by factoring $\tilde{\mathscr{H}}$ by the equivalence relation $\sim$, which quotients out $|\psi\rangle \in \tilde{\mathscr{H}}$, where either: (i) $|\psi\rangle$ is the state that would be obtained by restricting a state $|\varphi\rangle$, defined in terms of a graph Γ, to a state defined on a subgraph $\Gamma' \subset \Gamma$; or (ii) $|\psi\rangle$ is mapped into some other state $|\varphi\rangle$ by the action of the discrete group of automorphisms of Γ (i.e. the group of maps from links to links and from nodes to nodes, such that source and target relations are preserved). This is the Hilbert space of LQG (Rovelli, 2010, p. 2).

Spin Networks and Spinfoams

Having defined the Hilbert space of LQG, we must now say more regarding the states in this space. To do so, we first introduce the notion of a *spin network state*, before generalizing to *spinfoams*. A basis of $\mathscr{H}$ is labeled by three quantum numbers: (i) an abstract graph Γ (up to its isomorphisms), of the kind introduced above; (ii) a 'coloring' j_l of the links of Γ with irreducible representations of $SU(2)$; and (iii) a 'coloring' i_n of each node of Γ with $SU(2)$ representations ('intertwiners') (Wüthrich, 2017, p. 313). The states $|\Gamma, j_l, i_n\rangle$ labeled by these quantum numbers are called *spin network states*. The spin network states $|\Gamma, j_l, i_n\rangle$ are eigenstates of so-called area and volume operators defined on $\mathscr{H}$ (Rovelli, 2010, pp. 2–3); since these operators have discrete spectra (Rovelli, 2010; Wüthrich, 2017), one infers that spin network states represent 'granular' spacetime.

[95] Rovelli and Vidotto (2015) write $\mathscr{H}_\Gamma := L^2\left[SU(2)^L/SU(2)^N\right]$; strictly speaking this does not make mathematical sense, so I present the full structure of the Hilbert space above.

[96] For further details on measure theory, see (Schuller, 2016); a classic text is (Halmos, 1974).

[97] Again, see (Schuller, 2016).

[98] Let $\langle G, \tau\rangle$ be a locally compact Hausdorff topological group. If g is an element of G and S a subset of G, then we define the *left translate* of S under g as $gS := \{g \cdot s | s \in S\}$; similarly, we define the *right translate* of S under g as $Sg := \{s \cdot g | s \in S\}$. For Haar's original paper, see (Haar, 1933); for more recent discussion, see (Dunkl and Ramirez, 1971).

Roughly speaking, the ontology of LQG consists in quantum superpositions of spin network states (Rovelli, 2004; Rovelli and Vidotto, 2015). In order to be more precise about this, however, we must first introduce the notion of a *spinfoam*. Consider a spin network state $|\Gamma, j_l, i_n\rangle$, and imagine the representandum of this state evolving in an extra dimension ('time'). The links of the graph will sweep faces, and the nodes will sweep edges. This creates a *2-complex* with edges and faces, which describes the *evolution* of the spin network with respect to the newly introduced, 'temporal' parameter. The vertices of this 2-complex represent points at which the graph Γ changes; at these points, an edge splits into several edges, or vice versa, leading to new faces, and so to a different graph. The 'coloring' of a 2-complex with representations and intertwiners is called a *spinfoam* (Livine, 2010, p. 30).

Dynamics

I turn now to quantum superpositions of spinfoams, and to associating with these transition amplitudes (cf. §5.1). We begin by seeking to construct the amplitude associated with a particular 2-complex, taking this to be the product of amplitudes associated to the vertices, edges, and faces of the spinfoam, and dependent only upon local representations and intertwiners. Consider a given 2-complex Δ with boundary $\partial\Delta$ and a spin network state which defines the boundary state on $\partial\Delta$. The amplitude associated to Δ is a sum over all possible representations and intertwiners in the bulk and consistent with the boundary spin network: (Livine, 2010, p. 32)

$$\mathscr{A}\left[\Delta, \psi_{\partial\Delta}\right] = \sum_{j_f, i_e} \prod_f \mathscr{A}_f\left[j_f\right] \prod_e \mathscr{A}_e\left[i_e, j_{f\ni e}\right] \prod_\sigma \mathscr{A}_\sigma\left[j_{f\ni\sigma} i_{e\ni\sigma}\right]. \tag{5.4.4}$$

Here, the j_f and i_e for faces and edges on the boundary $f, e \in \partial\Delta$ are fixed and given by the choice of boundary state $\psi_{\partial\Delta}$. The amplitude $\mathscr{A}\left[\Delta, \psi_{\partial\Delta}\right]$ defines a discrete spinfoam path integral (cf. §5.1). The face weight $\mathscr{A}_f$ and edge weight $\mathscr{A}_e$ are considered to be kinematical and to define the path integral measure, while the amplitude $\mathscr{A}_\sigma$ associated to the 2-complex vertices σ contains all the dynamical information of the spinfoam model (Livine, 2010, p. 32).

The final step of the spinfoam approach to computing LQG transition amplitudes is to sum over all possible 2-complexes compatible with the boundary spin network state:

$$\mathscr{A}\left[\Gamma, \psi_\Gamma\right] = \sum_{\Delta|\partial\Delta=\Gamma} w(\Delta)\, \mathscr{A}\left[\Delta, \psi_\Gamma\right] \tag{5.4.5}$$

Here, $w(\Delta)$ is a statistical weight depending only upon the 2-complex Δ (and not on the representations and intertwiners living on it and defining the spinfoam), which depends on its symmetries (Livine, 2010, p. 33). As Livine states,

> This sum [(5.4.5)] is much less controlled by [*sic*—than] the previous sum over representations and intertwiners for a fixed 2-complex. It is usually ill-defined, it naively diverges [*sic*] and requires proper gauge-fixing. The standard way to define it non-perturbatively is through the group field theory formalism, which uses generalized matrix models to generate the sum over suitable 2-complexes...
> (Livine, 2010, p. 33)

For details on *group field theory*, see e.g. (Oriti, 2009, 2017); though philosophically interesting in its own right, I do not discuss further this theory in the remainder of this chapter.[99]

5.4.2 Background Independence

The above presentation of LQG constitutes sufficient background to embark upon a discussion of the background independence of this theory. In this regard, however, an obvious issue now arises: as is evident from the formalism presented above, LQG is not a field theory. Rather, KPMs are picked out by boundary spin network states $|\Gamma, j_l, i_n\rangle$; DPMs are those KPMs associated to which are quantum mechanical transition amplitudes consistent with the construction presented in §5.4.1. (In effect, this means that the only structure needed to pick out a KPM of LQG is a certain boundary condition—in other words, unlike the theories considered up to this point, the KPMs of LQG *just are* the theory's BPMs.) This means that (the quantum generalizations of—cf. §5.1) certain definitions of background independence, in particular Def. 6 (in terms of the absence of formulations featuring fixed fields) and Def. 9 (in terms of the absence of absolute fields), are not obviously applicable—since these definitions make particular reference to the presence/absence of fields in a given theory.[100] Moreover, the whole machinery of corner charges is adumbrated in the context of field theory, so—although there is perhaps work to be done here (in particular, to generalize the corner charge approach to discrete cases such as that encountered in LQG)—it is not that case that Def. 15 or Def. 19 can be applied in any straightforward manner to the case of LQG.

[99] The abstract, group-theoretic structure of group field theory raises further questions regarding how classical spacetime could 'emerge' from such a quantum theory of gravity—cf. footnote 94. For philosophical discussion of these matters (what Oriti calls the issue of *geometrogenesis* (Oriti, 2014, p. 186)), see (Oriti, 2014; Huggett, 2021).

[100] Of course, one could argue that, since it is not a field theory, LQG trivially *satisfies* Def. 6 and Def. 9. Such a response is unsatisfying, however, for it would open the door to *any* theory which is not a field theory qualifying as background independent. But should e.g. Newtonian particle mechanics, or the 'theory' of the singleton set {1}, with no dynamics, be considered background independent? Intuitively, no!

On the other hand, other definitions of background independence *are* applicable to LQG. Consider first Def. 5 (in terms of the absence of absolute objects); unlike GR, and essentially every other theory considered up to this point, it does not appear that LQG has absolute objects in its formulation—there is no structure in the theory which is fixed in all its DPMs. Similarly, considering now Belot's approach, taking the degrees of freedom in each KPM of the theory (which pick out a boundary spin network in each case) to be the geometrical degrees of freedom (a reasonable assumption, since these are the only degrees of freedom on offer), since it is the case that these degrees of freedom differ in each model, and (at least naïvely, absent some further argument) each of these models represents a *distinct* LQG quantum gravity model, it appears that Def. 12 (of full background independence) is satisfied.[101]

Finally, consider the quantum mechanical extension of Def. 10 (in terms of variational principles), introduced in §5.1—or, more directly, Def. 18 (i.e. the definition of background independence offered in terms of the path integral). The relevant objects to consider here are (5.4.4) and (5.4.5). Since there are no non-variational fields in either $\mathscr{A}_f$ or $\mathscr{A}_e$ (which fix the path integral measure), or $\mathscr{A}_\sigma$ (which fixes the dynamics, in analogy with the exponential of a classical action—cf. §5.1, and (Rovelli, 2001, ch. 7)), Def. 10 and Def. 18 do appear to be satisfied in this case.

5.4.3 Discussion

I return now to (1) and (2), raised at the beginning of this section in response to Matsubara's statements regarding the background independence of LQG (Matsubara, 2017, p. 3). Though I have argued that LQG *does* qualify as background independent on all (applicable) definitions of this notion (in particular, Def. 5, Def. 12, and Def. 10), I contest any claim that the theory is 'manifestly' background independent—rather, verdicts regarding the background independence of LQG may only be issued once one has in hand: (a) a clear picture of the structure of the theory; and (b) precise definitions of background independence, which can be applied to this theory. It is plausible that such claims regarding the background independence of LQG are often issued on the grounds that the theory is—in a sense that is not true of e.g. perturbative string theory—a (more-or-less) direct quantization of GR, which is (on most accounts) a background independent theory itself. However, such arguments elide the fact that, having constructed LQG, it is now a *distinct* theory, and should be evaluated *on its own terms*.

[101] It's not so obvious that this verdict will continue to hold when one couples LQG to matter—on which, see (Rovelli and Vidotto, 2015, ch. 9).

What of (2), which regards a comparative judgment of the background independence of LQG and perturbative string theory? Here the situation is more subtle than one may at first appreciate, for, on several approaches to background independence, perturbative string theory *does* qualify as background independent (see §5.2).[102] Though there are (as we have seen) several subtleties regarding this verdict in the context of string theory (e.g. regarding whether different embeddings of the string worldsheet in target space should be considered 'gauge equivalent'—see §5.2.3)—to an extent which, broadly speaking, does not arise in the context of LQG—the fact is that, on many accounts and certain plausible understandings of string theory, *both* theories qualify as background independent. Consequently, any arguments on the part of advocates of LQG that string theory is unviable in virtue of its (alleged) lack of background independence (see e.g. (Smolin, 2006, 2008)) are, at the very least, questionable.

Finally, it is worth addressing the question: why, in the literature, does the notion of background independence play such a significant role in the presented motivations for LQG? (That this is so is undeniable: Smolin has written a book on the topic (Smolin, 2008), and even in technical introductions to the theory, e.g. (Ashtekar and Lewandowski, 2004; Rovelli, 2004; Rovelli and Vidotto, 2015), one finds strong reference to the purported background independence of this theory.) I take it that the answer to this question is this: unlike proponents of other approaches to quantum gravity (e.g. string theory), those working on LQG take background independence to be an important *guiding principle* which can aid one in the development of a quantum theory of gravity; thus, it is of no surprise that this principle is affording greater significance in these communities than in others (see, for example, (Rovelli, 2004, p. 9), in which background independence is explicitly cited as such a guiding principle). This is not to say that proponents of LQG need be tied to any *specific* definition of background independence, such as those presented in this book. (Indeed, as we have already seen, it is not always the case that a given classical notion of background independence is applicable to a certain quantum theory.) Rather, the thought (I take it) is that the *spirit* of background independence should live on in a quantum theory of gravity. The foregoing discussion in this chapter should make clear that I concur with proponents of LQG that this theory of quantum gravity makes good on this ambition.

That said, I do not wish to endorse a stronger view, evinced in the writings of e.g. Smolin (see again e.g. (Smolin, 2006, 2008)), that background independence is not merely a guiding principle for theory construction, but is, moreover, a *rationalist motivation* (to use terminology introduced by Read and Le Bihan (2021)) which *must* be fulfilled in any quantum theory of gravity. The first reason

[102] Altough necessarily it does not on Rovelli's characterization of background independence presented at (Rovelli, 2004, p. 9).

for which I reject this claim is that, as pointed out, not all extant definitions of background independence are appropriate in the quantum regime, or for theories the kinematical structure of which does not consist of fields on a differentiable manifold. Thus, to insist on principles developed in classical contexts would seem to me to be an unnecessary, and potentially problematic, straightjacketing of our ambitions in the context of the development of new physical theories. (On the other hand, to insist that the 'spirit' of background independence be not just a guiding principle, but a rationalist motivation in the above sense, is arguably problematically vague.) Second, I concur with the classic view of van Fraassen (1980) that the ultimate arbiter of the success of a scientific theory is whether it 'saves the phenomena'—that is, whether it is empirically adequate. If a theory is able to save the phenomena while nevertheless violating one definition of background independence or another, then so much the worse for that definition. Smolin fairly evidently has a stricter understanding of what it would take for a (final) scientific theory to be 'successful', according to which *everything must be explained*. Setting aside philosophical questions regarding the notion of 'explanation' at play here,[103] the thought might be taken to be that if one's theory contains a piece of fixed structure, then there should be some explanation as to why that structure is fixed; if everything is ultimately to be explained, then in the final theory of physics there can be no such fixed structures. If this is correct as an understanding of Smolin, then, in my view, the point is indeed best made in terms of (the desired sense of) scientific explanation, rather than in terms of background independence, given the flexibility and potential lack of univocity in the latter notion by now witnessed in abundance. Moreover, even doing so, an author with a more empiricist outlook (such as myself—again, cf. (Read and Le Bihan, 2021)) might reject Smolin's stricter criteria for the success of a scientific theory.

[103] There are many models of scientific explanation, from the (in)famous 'deductive-nomological model' (Hempel and Oppenheim, 1948), to causal models (Lewis, 1986), to unificatory models (Friedman, 1974; Kitcher, 1989). For a recent collection on non-causal models of explanation in science, see (Reutlinger and Saatsi, 2018).

6
Conclusions

Over the course of this book, I have worked with six central classical definitions of background independence: Def. 5 (in terms of the absence of absolute objects), Def. 6 (in terms of the non-existence of theory formulations featuring fixed fields), Def. 9 (in terms of the absence of absolute fields), Def. 10 (in terms of variational principles), Def. 12 (in terms of a coincidence of geometrical and physical degrees of freedom in a theory), and Def. 15 (in terms of the existence of non-trivial corner charges). In addition, I have presented (i) certain quantum extensions of these definitions—*viz.*, Def. 18 (the version of Def. 10 suitable to path integral formulations of physical theories) and Def. 19 (the version of Def. 15 suitable—again—to path integral formulations of physical theories); and (ii) approaches to background independence developed in the context of particular theories, e.g. Def. 16 and Def. 17 (presented in my discussions of Machian relationism), and Def. 20 (presented in the context of holography).

One of my central purposes in availing myself of so many definitions of background independence has been to provide a roadmap (of sorts) around an otherwise extremely complex and convoluted literature: I hope that this book can be helpful in (a) navigating these multifarious definitions and their interrelations; (b) appreciating the contexts in which a certain definition is applicable and other contexts in which the same definition is not applicable; and (c) recognizing features of approaches to background independence which have not been widely or systematically appreciated in the literature up to this point (for example, the fact that while most approaches to background independence construe the notion formally, Def. 12—Belot's proposal—and Def. 17—Gryb's proposal—buck the trend by having an important interpretative aspect[1]).

So: bringing clarity and order to the literature on background independence has been one goal. But a second—equally important—goal has been likewise to bring clarity and order to our assessments of background independence. All too often, one finds authors making declarations of background independence 'from the hip': this is, perhaps, most obvious in the quantum gravity context, in the widespread adjudications (often without substantial argument) that string theory is *not* background independent, and that LQG *is* background independent. We should strive to do better: giving precise definitions of background independence,

[1] Cf. interpretative versus formal approaches to symmetries, as discussed by Dasgupta (2016).

Background Independence in Classical and Quantum Gravity. James Read, Oxford University Press.
© James Read 2023. DOI: 10.1093/oso/9780192889119.003.0006

and laying them against the precise details of the particular theories of interest. Although the verdicts issued are—as we have seen time and time again in this piece—accordingly more nuanced, it is only in this way that we can move beyond the mere trading of intuitions in these debates.

I have also argued that there are ancillary benefits to this more precise and systematic methodology which cut in two directions. First: consideration of some theories, e.g. NCT, allows one to identify and amend obvious flaws in some approaches to background independence: we saw this particularly clearly in §3.6 in the case of Belot's approach.[2] But second: applying these precise definitions of background independence can also help us to come to appreciate better the formal structure of our theories: to take just one example, by identifying hitherto-unappreciated absolute objects in GR (see §3.3). This I take to be one of the main intellectual payoffs to be had from the study of background independence.

Returning to the topic of intuitions: one might worry that the methodology followed in this book is, even in spite of the above critiques of 'intuition-mongering' in the foundations of physics, far from devoid of any appeal to intuitions (they are, as Knox (2011) might say, "waiting in the wings"). But I would like to argue that the role of intuitions in this work has been very different to the role of intuitions in many discussions of background independence up to this point. In this work, I (following other authors who have worked on this topic) have used *a priori* intuitions to guide the construction of precise definitions of background independence, which I have then laid against various physical theories in order to assess their background independence. Transparently, this is very different to using intuitions to adjudicate directly on whether a given theory is, or is not, background independent (often without even writing down the formal structure of that theory!).

Is the use of intuitions in this work in this way still appropriately 'naturalistic'? My answer to this question is *yes*: intuitions have motivated an array of definitions of background independence, each of which has been useful and valuable in illuminating the conceptual structure of a range of physical theories, both classical and quantum. But I make no claim that there is one univocal notion of 'background independence' which it is the philosopher of physics' job to discover (compare—perhaps pejoratively—the project of analyzing the concept of 'knowledge' in epistemology): rather, there is a plurality of definitions, each of which is helpful in enriching our understanding of the workings of our physical theories: and it is this enrichment, rather than identifying any 'true' notion of background independence, which is the end goal. Insofar as these notions are, in my view, best

[2] One is reminded here of Lakatos' Hegelian approach to mathematical proofs: see (Lakatos, 1976), and (Musgrave and Pigden, 2016, §3.1) for discussion.

construed as tools for the *study* of physical theories, I maintain that the approach advocated in this work is naturalistic.[3]

In light of my 'pluralist' approach to background independence (cf. pluralism about symmetries in physics which, in the form presented by Read and Møller-Nielsen (2020b), I also take to be suitably naturalistic), it should be clear that I do not take it that there is *one* notion of background independence which *must* be manifest in future physical theorizing. Various definitions of background independence can be used as *guiding principles* for the construction of new physical theories (see, in this regard, (Crowther, 2021)), and, as we have seen in §5.4, the more nebulous 'spirit' of background independence, loosely construed, can also be used as such a guiding principle (this I take to be close to the thinking of Rovelli—see (Rovelli, 2004, ch. 1)); however, my stance is again sufficiently naturalistic (cf. (van Fraassen, 2002)) that I would never wish to understand background independence as a *rationalist motivation* (Read and Le Bihan, 2021) which *must* be manifest in any future physical theorizing. In addition, it should be clear that I reject any thought that there is some privileged property of 'background independence' which sets GR apart from other spacetime theories: as we have seen, the story is much more delicate, with GR satisfying many (but not all) of our plurality of definitions, but this likewise being the case for many other spacetime theories.

In sum, then: as I see it, the central benefits of studying background independence in the systematic manner presented in this book are two. First: doing so affords us a more developed understanding of the structure of various physical theories, and their interrelations (thereby making good on the project, proposed by Lehmkuhl (2017), of exploring in a systematic manner the 'space of spacetime theories'—for earlier espousal of such a project, see (Thorne et al., 1973)): the importance of this in the foundations of physics surely needs no argument. Second: doing so affords us a rich array of potential 'guiding principles' (none uniquely privileged) for the development of future physical theories.

I hope that this work has been sufficiently systematic and far-reaching to act as a guide in future explorations of background independence. There remains much to be done: for example, recently a concern has been raised regarding 'asymptotic safety' approaches to quantum gravity—which attempt to secure a theory of quantum gravity via considerations of renormalization group flow; the details need not detain us here, but for background, see e.g. (Percacci, 2009)—that they are not background independent (Reuter and Weyer, 2009). (Roughly, the concern is the same as that with perturbative string theory: one must presuppose some fixed metrical structure in order to proceed with the approach.) It should be fairly straightforward to apply the lessons of this book to asymptotic safety, and

[3] There is a substantial literature on the role of intuitions in analytic philosophy: see (Pust, 2017, §2) and (Williamson, 2007, ch. 7) and references therein for discussion.

to other approaches to quantum gravity which may be developed in the future, in order to address such concerns precisely. First: one can ask: given the definitions presented in this book, is it indeed true that this approach to quantum gravity is background dependent? Second: even if this *is* true, is that any reason to reject the approach, given that (at least as I have argued!) background independence is a guiding principle rather than a rationalist motivation—an heuristic rather than a norm?

Bibliography

Aldrovandi, R. and Pereira, J. G. (2013). *Teleparallel Gravity: An Introduction*, Springer, Berlin.

Ammon, M. and Erdmenger, J. (2015). *Gauge/Gravity Duality: Foundations and Applications*, Cambridge University Press, Cambridge.

Anderson, J. L. (1964). Relativity principles and the role of coordinates in physics, *in* H.-Y. Chiu and W. F. Hoffmann (eds), *Gravitation and Relativity*, W. A. Benjamin, New York, pp. 175–194.

Anderson, J. L. (1967). *Principles of Relativity Physics*, Academic Press, New York.

Anderson, J. L. and Gautreau, R. (1969). Operational formulation of the equivalence principle, *Physical Review* **185**(5): 1656–1661.

Ashtekar, A. (2009a). Loop quantum cosmology: An overview, *General Relativity and Gravitation* **41**(4): 707–741.

Ashtekar, A. (2009b). Singularity resolution in loop quantum cosmology: A brief overview, *Journal of Physics: Conference Series* **189**.

Ashtekar, A., Corichi, A., and Singh, P. (2008). Robustness of key features of loop quantum cosmology, *Physical Review D* **77**(2): 024046.

Ashtekar, A. and Lewandowski, J. (2004). Background independent quantum gravity: A status report, *Classical and Quantum Gravity* **21**: R53–R152.

Bailin, D. and Love, A. (1987). Kaluza-Klein theories, *Reports on Progress in Physics* **50**: 1087–1170.

Bain, J. (2004). Theories of Newtonian gravity and empirical indistinguishability, *Studies in the History and Philosophy of Modern Physics* **35**: 345–376.

Baker, D. J. (2021). Knox's inertial spacetime functionalism (and a better alternative), *Synthese*. **199**(2): 277–298. **URL:** https://doi.org/10.1007/s11229-020-02598-z

Barbour, J. (1994). The timelessness of quantum gravity. Vol. I: The evidence from the classical theory, *Classical and Quantum Gravity* **11**: 2853–2873.

Barbour, J. (1999). *The End of Time: The Next Revolution in Our Understanding of the Universe*, Weidenfeld & Nicholson, London.

Barbour, J. (2003). Scale-invariant gravity: Particle dynamics, *Classical and Quantum Gravity* **20**: 1543–1570.

Barbour, J. (2008). The nature of time. **URL:** https://arxiv.org/abs/0903.3489

Barbour, J. (2010). Shape dynamics: An introduction, *Proceedings of the Conference Quantum Field Theory and Gravity*, Regensburg.

Barbour, J. (2015). Does time differ from change? Philosophical appraisal of the problem of time in quantum gravity and in physics: A response, *Studies in the History and Philosophy of Modern Physics* **52**: 55–61.

Barbour, J. and Bertotti, B. (1982). Mach's principle and the structure of dynamical theories, *Proceedings of the Royal Society, London A* **382**: 295–306.

Barbour, J., Foster, B., and 'O Murchadha, N. (2002). Relativity without relativity, *Classical and Quantum Gravity* **19**: 3217–3248.

Baron, S., Bihan, B. L., and Read, J. (2022). Science and possibility. Unpublished manuscript.

Becker, K., Becker, M., and Schwarz, J. (2007). *String Theory and M-Theory: A Modern Introduction*, Cambridge University Press, Cambridge.

Bekaert, X. and Morand, K. (2016). Connections and dynamical trajectories in generalised Newton-Cartan gravity. Vol. I: An intrinsic view, *Journal of Mathematical Physics* **57**: 022507.

Belot, G. (2008). An elementary notion of gauge equivalence, *General Relativity and Gravitation* **40**: 199–215.

Belot, G. (2011). Background-independence, *General Relativity and Gravitation* **43**: 2865–2884.

Belot, G. (2013). Symmetry and equivalence, *in* R. Batterman (ed.), *The Oxford Handbook of Philosophy of Physics*, Oxford University Press, Oxford, pp. 318–339.

Belot, G. (2018). Fifty million Elvis fans can't be wrong, *Noûs* **52**(4): 946–981.

Blumenhagen, R., Lüst, D., and Theisen, S. (2013). *Basic Concepts of String Theory*, Springer Series in Theoretical and Mathematical Physics, Springer, Berlin.

Bojowald, M. and Brahma, S. (2017). Signature change in loop quantum gravity: Two-dimensional midisuperspace models and dilaton gravity, *Physical Review D* **95**: 124014.

Bojowald, M. and Mielczarek, J. (2015). Some implications of signature-change in cosmological models of loop quantum gravity, *Journal of Cosmology and Astroparticle Physics* **8**: 52.

Brading, K. (2005). A note on general relativity, energy conservation, and Noether's theorems, *in* A. J. Kox and J. Eisenstaedt (eds), *The Universe of General Relativity*. Vol. 11: of *Einstein Studies*, Birkhäuser, Basel, pp. 125–135.

Brown, H. R. (2005). *Physical Relativity: Space-Time Structure from a Dynamical Perspective*, Oxford University Press, Oxford.

Brown, H. R. and Lehmkuhl, D. (2015). Einstein, the reality of space, and the action-reaction principle, *in* P. Ghose (ed.), *The Nature of Reality*, Pickering and Chatto, London.

Brown, H. R. and Pooley, O. (2001). The origins of the spacetime metric: Bell's Lorentzian pedagogy and its significance in general relativity, *in* C. Callender and N. Huggett (eds), *Physics Meets Philosophy at the Plank Scale*, Cambridge University Press, Cambridge.

Brown, H. R. and Pooley, O. (2006). Minkowski space-time: A glorious non-entity, *in* D. Dieks (ed.), *The Ontology of Spacetime*, Elsevier, London.

Brown, H. R. and Read, J. (2021). The dynamical approach to spacetime, *in* E. Knox and A. Wilson (eds), *The Routledge Companion to Philosophy of Physics*, Routledge, London.

Butterfield, J. and Gomes, H. (2021a). Functionalism as a species of reduction, *in* C. Soto (ed.), *Current Debates in the Philosophy of Science: In Honor of Roberto Torretti*, Springer, New York.

Butterfield, J. and Gomes, H. (2021b). Geometrodynamics as functionalism about time, *in* C. Kiefer (ed.), *From Quantum to Classical: Essays in Memory of Dieter Zeh*, Springer, New York. Forthcoming.

Callan, C. G., Friedan, D., Martinec, E. J., and Perry, M. J. (1985). Strings in background fields, *Nuclear Physics B* **262**(4): 593–609.

Caulton, A. (2015). The role of symmetry in the interpretation of physical theories, *Studies in History and Philosophy of Modern Physics* **52**: 153–162.

Christian, J. (1997). Exactly soluble sector of quantum gravity, *Physical Review D* **56**(8): 4844–4877.

Ciambelli, L. (2022). From asymptotic symmetries to the corner proposal. **URL**: https://arxiv.org/abs/2212.13644

Claus, P., Kallosh, R., and Van Proeyen, A. (1998). M5-brane and superconformal $(0,2)$ tensor multiplet in six dimensions, *Nuclear Physics B* **518**: 117–150.

Crowther, K. (2016). *Effective Spacetime Geometry: Understanding Emergence in Effective Field Theory and Quantum Gravity*, Springer, Cham.

Crowther, K. (2019). When do we stop digging? Conditions on a fundamental theory of physics, *in* A. Aguirre, B. Foster, and Z. Merali (eds), *What is Fundamental?*, Springer, New York, pp. 123–133.

Crowther, K. (2021). Defining a crisis: The roles of principles in the search for a theory of quantum gravity, *Synthese* **198**(14), 3489–3516. **URL**: https://doi.org/10.1007/s11229-018-01970-4.

Curiel, E. (2016). Kinematics, dynamics, and the structure of physical theory. **URL**: https://arxiv.org/abs/1603.02999

Curiel, E. (2019). On geometric objects, the non-existence of a gravitational stress-energy tensor, and the uniqueness of the Einstein field equation, *Studies in History and Philosophy of Modern Physics* **66**: 90–102.

Dasgupta, S. (2016). Symmetry as an epistemic notion (twice over), *British Journal for the Philosophy of Science* **67**(3): 837–878.

de Andrade, V., Guillen, L., and Pereira, J. (2000). Teleparallel gravity: An overview. **URL**: https://www.worldscientific.com/doi/abs/10.1142/9789812777386_0144

de Haro, S. (2016). Reply to James Read on background-independence. **URL**: https://beyondspacetimedotnet.files.wordpress.com/2016/06/de-haro-read.pdf

de Haro, S. (2017a). Dualities and emergent gravity: Gauge/gravity duality, *Studies in History and Philosophy of Modern Physics* **59**: 109–125.

de Haro, S. (2017b). The invisibility of diffeomorphisms, *Foundations of Physics* **47**: 1464–1497.

de Haro, S., Mayerson, D. R., and Butterfield, J. (2016). Conceptual aspects of gauge/gravity duality, *Foundations of Physics* **46**: 1381–1425.

de Haro, S., Skenderis, K., and Solodukhin, S. N. (2001). Holographic reconstruction of spacetime and renormalization in the ADS/CFT correspondence, *Communications in Mathematical Physics* **217**: 595–622.

de Haro, S., Teh, N., and Butterfield, J. (2017). Comparing dualities and gauge symmetries, *Studies in History and Philosophy of Modern Physics* **59**: 68–80.

Deutsch, D. and Hayden, P. (2000). Information flow in entangled quantum systems, *Proceedings of the Royal Society of London A* **456**: 1759–1774.

Dewar, N. (2015). Symmetries and the philosophy of language, *Studies in the History and Philosophy of Modern Physics* **52**: 317–327.

Dewar, N. (2018). Maxwell gravitation, *Philosophy of Science* **85**: 249–270.

Dewar, N. (2019). Sophistication about symmetries, *British Journal for the Philosophy of Science* **70**: 485–521.

Dewar, N. and Read, J. (2020). Conformal invariance of the Newtonian Weyl tensor, *Foundations of Physics* **50**: 1418–1425.

Dewar, N. and Weatherall, J. O. (2018). On gravitational energy in Newtonian theories, *Foundations of Physics* **48**: 558–578.

Dieks, D., van Dongen, J., and de Haro, S. (2015). Emergence in holographic scenarios for gravity, *Studies in History and Philosophy of Modern Physics* **52**: 203–216.

Dougherty, J. (2019). Review of *The Emergence of Spacetime in String Theory* by Tiziana Vistarini, *Notre Dame Philosophical Reviews*. **URL**: https://ndpr.nd.edu/reviews/the-emergence-of-spacetime-in-string-theory/

Dougherty, J. (2020). Large gauge transformations and the strong CP problem, *Studies in History and Philosophy of Modern Physics* **69**: 50–66.

Duerr, P. M. (2021). Against 'functional gravitational energy': A critical note on functionalism, selective realism, and geometric objects and gravitational energy, *Synthese* **199**(Suppl 2): 299–333.

Dunkl, C. F. and Ramirez, D. E. (1971). *Topics in Harmonic Analysis*, Appleton-Century-Crofts, London.

Dürr, P. and Read, J. (2019). Gravitational energy in Newtonian gravity: A response to Dewar and Weatherall, *Foundations of Physics* **49**: 1086–1110.

Dürr, P. and Read, J. (2023). Reconsidering conventionalism: An invitation to a sophisticated philosophy for modern (space-)times. **URL**: http://philsci-archive.pitt.edu/22172/

Earman, J. (1989). *World Enough and Space-Time: Absolute Versus Relational Theories of Space and Time*, MIT Press, Cambridge, MA.

Earman, J. (2003). The cosmological constant, the fate of the universe, unimodular gravity, and all that, *Studies in History and Philosophy of Science* **34**: 559–577.

Earman, J. (2006). Two challenges to the requirement of substantive general covariance, *Synthese* **148**(2): 443–468.

Earman, J. and Norton, J. (1987). What price spacetime substantivalism? The hole story, *British Journal for the Philosophy of Science* **38**(4): 515–525.

Ehlers, J. (1981). Über den newtonschen grenzwert der einsteinschen gravitationstheorie, *in* J. Nitsch, J. Pfarr, and E. W. Stachow (eds), *Grundlagen Probleme der Modernen Physik*, Mannheim: Bibliographisches Institut, pp. 65–84.

Erbin, H. (2021). *String Field Theory*, Springer International Publishing, New York.

Fefferman, C. and Graham, C. R. (1985). Conformal invariants, *Elie Cartan et les Mathématiques d'aujourd'hui*, Astérisque, Paris, pp. 95–107.

Ferrari, J. A., Griego, J., and Falco, E. E. (1989). The Kaluza-Klein theory and four-dimensional spacetime, *General Relativity and Gravitation* **21**: 69–78.

Feynman, R. P., Morinigo, F. B., and Wagner, W. G. (1995). *Feynman Lectures on Gravitation*, Addison-Wesley, Reading, MA.

Fine, D. and Fine, A. (1997). Gauge theory, anomalies and global geometry: The interplay of physics and mathematics, *Studies in History and Philosophy of Modern Physics* **28**(3): 307–323.

François, J. (2019). Artificial versus substantial gauge symmetries: A criterion and an application to the electroweak model, *Philosophy of Science* **86**(3): 472–496.

Freidel, L., Geiller, M., and Pranzetti, D. (2020). Edge modes of gravity. Part I. Corner potentials and charges, *Journal of High Energy Physics* **2020**(26): 1–52.

Freidel, L. and Teh, N. J. (2022). Substantive general covariance and the Einstein-Klein dispute: A Noetherian approach, *in* J. Read and N. J. Teh (eds), *The Philosophy and Physics of Noether's Theorems*, Cambridge University Press, Cambridge, pp. 274–295.

Friedman, M. (1974). Explanation and scientific understanding, *Journal of Philosophy* **71**(1): 5–19.

Friedman, M. (1983). *Foundations of Space-Time Theories: Relativistic Physics and the Philosophy of Science*, Princeton University Press, Princeton.

Geracie, M., Prabhu, K., and Roberts, M. M. (2015). Curved non-relativistic spacetimes, Newtonian gravitation and massive matter, *Journal of Mathematical Physics* **56**(10): 103505. **URL**:https://doi.org/10.1063/1.4932967

Gibbons, G. W. and Hartle, J. B. (1990). Real tunnelling geometries and the large-scale topology of the universe, *Physical Review D* **42**(8): 2458–2468.

Gielen, S., de León Ardón, R., and Percacci, R. (2018). Gravity with more or less gauging, *Classical and Quantum Gravity* **35**(19): 195009.

Giulini, D. (2007). Remarks on the notions of general covariance and background independence, *Lecture Notes in Physics* **721**: 105–120.

Gomes, H. (2011). Classical gauge theory in riem, *Journal of Mathematical Physics* **52**(8): 082501. **URL**: https://doi.org/10.1063/1.3603990

Gomes, H. (2013). Poincaré invariance and asymptotic flatness in shape dynamics, *Phys. Rev. D* **88**: 024047. **URL**:https://link.aps.org/doi/10.1103/PhysRevD.88.024047

Gomes, H. (2011). Classical gauge theory in riem, *Journal of Mathematical Physics* **52**(8): 082501. **URL**:https://doi.org/10.1063/1.3603990

Gomes, H. and Gryb, S. (2021). Angular momentum without rotation: Turbocharging relationalism, *Studies in History and Philosophy of Science* **88**: 138–155. **URL**: https://www.sciencedirect.com/science/article/pii/S0039368121000704

Gomes, H. and Koslowski, T. (2012). The link between general relativity and shape dynamics, *Classical and Quantum Gravity* **29**: 075009.

Gomes, H. and Riello, A. (2020). Large gauge transformations, gauge invariance, and the QCD θ_{YM}-term. **URL**: https://arxiv.org/abs/2007.04013

Gourgoulhon, É. (2007). 3+1 Formalism and bases of numerical relativity, Lecture notes.

Greaves, H. and Wallace, D. (2014). Empirical consequences of symmetries, *British Journal for the Philosophy of Science* **65**: 59–89.

Green, M. B., Schwarz, J. H., and Witten, E. (1987). *Superstring Theory*, Vol. 1, Cambridge University Press, Cambridge.

Gryb, S. (2010). A definition of background independence, *Classical and Quantum Gravity* **27**(2): 215018.

Haar, A. (1933). Der Massbegriff in der Theorie der Kontinuierlichen Gruppen, *Annals of Mathematics* **2**(34): 147–169.

Halmos, P. R. (1974). *Measure Theory*, Springer, New York.

Halvorson, H. and Mueger, M. (2006). Algebraic quantum field theory, *in* J. Butterfield and J. Earman (eds), *Handbook of the Philosophy of Physics*, Kluwer Academic Publishers, Dordrecht.

Hammad, F., Dijamco, D., Torres-Rivas, A., and Bérubé, D. (2019). Noether charge and black hole entropy in teleparallel gravity, *Physical Review D* **100**(12): 124040.

Hansen, D., Hartong, J., and Obers, N. A. (2019a). Action principle for Newtonian gravity, *Physical Review Letters* **122**: 061106.

Hansen, D., Hartong, J., and Obers, N. A. (2019b). Gravity between Newton and Einstein, *International Journal of Modern Physics* **28**(14): 1944010.

Hansen, D., Hartong, J., and Obers, N. A. (2020). Non-relativistic gravity and its coupling to matter, *Journal of High Energy Physics*.

Hempel, C. and Oppenheim, P. (1948). Studies in the logic of explanation, *Philosophy of Science* **15**: 135–175.

Hiskes, A. L. D. (1984). Space-time theories and symmetry groups, *Foundations of Physics* **14**(4): 307–332.

Hoefer, C. (2000). Energy conservation in gtr, *Studies in the History and Philosophy of Modern Physics* **31**(2).

Horowitz, G. T. (1988). String theory without spacetime, *in* A. Ashtekar and J. Stachel (eds), *Conceptual Problems in Quantum Gravity*, Springer.

Huggett, N. (2016). The string theoretic explanation of general relativity. Unpublished manuscript.

Huggett, N. (2017). Target space $\neq$ space, *Studies in the History and Philosophy of Modern Physics* **59**: 81–88.

Huggett, N. (2021). Spacetime 'emergence', *in* E. Knox and A. Wilson (eds), *The Routledge Companion to Philosophy of Physics*, Routledge, London.

Huggett, N. and Vistarini, T. (2015). Deriving general relativity from string theory, *Philosophy of Science* **82**(5): 1163–1174.

Ismael, J. and van Fraassen, B. C. (2003). Symmetries as a guide to superfluous theoretical structure, *in* K. Brading and E. Castellani (eds), *Symmetries in Physics: Philosophical Reflections*, Cambridge University Press, Cambridge.

Jackson, J. D. (1999). *Classical Electrodynamics*, 3 edn, John Wiley & Sons, Inc, London.

Jacobs, C. (2021). *Gauge and Explanation: Are Gauge-Quantities Explanatory?*, PhD thesis, University of Oxford.

Kaluza, T. (1921). Zum unitätsproblem in der physik, *Sitzungsber. Preuss. Akad. Wiss. Berlin. Math. Phys.* **33**: 966–972.

Khesin, B. and Wendt, R. (2009). *The Geometry of Infinite-Dimensional Groups*, Springer, Berlin.

Kitcher, P. (1989). Explanatory unification and the causal structure of the world, *Minnesota Studies in the Philosophy of Science*, Vol. 13, University of Minnesota Press, Minnesota, pp. 410–503.

Klein, O. (1926). Quantentheorie und fünfdimensionale Relativitätstheorie, *Zeitschrift für Physik A* **37**: 895–906.

Knox, E. (2011). Newton-Cartan theory and teleparallel gravity: The force of a formulation, *Studies in the History and Philosophy of Modern Physics* **42**: 264–275.

Knox, E. (2013). Effective spacetime geometry, *Studies in History and Philosophy of Modern Physics* **44**: 346–356.

Knox, E. (2014). Newtonian spacetime structure in light of the equivalence principle, *British Journal for the Philosophy of Science* **65**(4): 863–880.

Knox, E. (2019). Physical relativity from a functionalist perspective, *Studies in History and Philosophy of Modern Physics* **67**: 118–124.

Kretschmann, E. (1917). Über den physikalischen Sinn der Relativitätspostulate, *Annalen der Physik* **53**: 575–614.

Kuchař, K. (1973). Canonical quantisation of gravity, *in* W. Israel (ed.), *Relativity, Astrophysics, and Cosmology*, D. Reidel, Dordrecht, pp. 237–288.

Kuchař, K. (1981). Canonical methods of quantisation, *in* C. J. Isham, R. Penrose, and D. W. Sciama (eds), *Quantum Gravity. Vol. 2: A Second Oxford Symposium*, Clarendon Press, Oxford, p. 329ff.

Künzle, H.-P. (1972). Galilei and Lorentz structures on space-time: Comparison of the corresponding geometry and physics, *Annales Institute Henri Poincaré* **17**: 337–362.

Künzle, H.-P. (1976). Covariant Newtonian limts of Lorentz space-times, *General Relativity and Gravitation* **7**: 445–457.

Ladyman, J., Ross, D., Spurrett, D., and Collier, J. (2007). *Every Thing Must Go: Metaphysics Naturalised*, Oxford University Press, Oxford.

Lakatos, I. (1976). *Proofs and Refutations: The Logic of Mathematical Discovery*, Cambridge University Press, Cambridge.

Lam, V. (2011). Gravitational and nongravitational energy: The need for background structures, *Philosophy of Science* **78**(5).

Le Bihan, B. (2018). Space emergence in contemporary physics: Why we do not need fundamentality, layers of reality and emergence, *Disputatio* **10**(49): 71–95.

Le Bihan, B. and Read, J. (2018). Duality and ontology, *Philosophy Compass* **13**: e12555.

Lee, J. and Wald, R. M. (1990). Local symmetries and constraints, *Journal of Mathematical Physics* **31**(3): 725–743. **URL**: https://doi.org/10.1063/1.528801

Lehmkuhl, D. (2009). Spacetime Matters: On Super-substantivalism, General Relativity, and Unified Field Theories, D.Phil. thesis, University of Oxford.

Lehmkuhl, D. (2011). Mass-energy-momentum: Only there because of spacetime?, *British Journal for the Philosophy of Science* **62**: 453–488.

Lehmkuhl, D. (2014). Why Einstein did not believe that general relativity geometrizes gravity, *Studies in History and Philosophy of Modern Physics* **46**: 316–326.

Lehmkuhl, D. (2017). *Introduction: Towards a Theory of Spacetime Theories*, Birkhäuser, Basel.

Levin, J. (2018). Functionalism, *in* E. N. Zalta (ed.), *The Stanford Encyclopedia of Philosophy* Stanford University Press, Stanford.

Lewis, D. (1986). Causal explanation, *Philosophical Papers*, Vol. 2, Oxford University Press, Oxford.

Linnemann, N. and Read, J. (2021a). Constructive axiomatics in spacetime physics. Part I: Walkthrough to the Ehlers-Pirani-Schild axiomatisation.

Linnemann, N. and Read, J. (2021b). On the status of Newtonian gravitational radiation, *Foundations of Physics* **51**(53).

Livine, E. (2010). *The Spinfoam Framework for Quantum Gravity*, PhD thesis, Ecole Normale Supérieure de Lyon.

Luc, J. (2022). Arguments from scientific practice in the debate about the physical equivalence of symmetry-related models, *Synthese* **200**(72): 1–29.

Malament, D. (2012). *Topics in the Foundations of General Relativity and Newtonian Gravitation Theory*, University of Chicago Press, Chicago.

Maldacena, J. (1998). The large *n* limit of superconformal field theories and supergravity, *Advances in Theoretical and Mathematical Physics* **2**: 231–252.

Maldacena, J. (2005). The illusion of gravity, *Scientific American* **293**(5):56–63.

Martens, N. C. M. and Read, J. (2020). Sophistry about symmetries?, *Synthese*. **199**: 315–344. **URL**: https://doi.org/10.1007/s11229-020-02658-4

Matsubara, K. (2017). Quantum gravity and the nature of space and time, *Philosophy Compass* **12**: e12405.

McLaughlin, B. (2006). Is role-functionalism committed to epiphenomenaliam?, *Consciousness Studies* **13**: 39–66.

Mercati, F. (2018). *Shape Dynamics: Relativity and Relationalism*, Oxford University Press, Oxford.

Misner, C., Thorne, K., and Wheeler, J. (1973). *Gravitation*, Freeman & Co., San Francisco.

Møller-Nielsen, T. (2017). Invariance, interpretation, and motivation, *Philosophy of Science* **84**(5): 1253–1264.

Morgado Patrício, A. (2013). 5D Kaluza-Klein theories: A brief review, https://fenix.tecnico.ulisboa.pt/downloadFile/3779580604342/kaluza.pdf.

Musgrave, A. and Pigden, C. (2016). Imre lakatos, *in* E. N. Zalta (ed.), *The Stanford Encyclopedia of Philosophy*, Metaphysics Research Lab, Stanford University Press, Stanford **URL**: https://plato.stanford.edu/archives/fall2016/entries/lakatos/

Newton, I. (2014). *Philosophical Writings*, revised edn, Cambridge University Press, Cambridge Press.

Nijenhuis, A. (1952). *Theory of the Geometric Object*, PhD thesis, University of Amsterdam.

Norton, J. D. (1989). Coordinates and covariance: Einstein's view of space-time and the modern view, *Foundations of Physics* **19**(12): 1215–1263.

Norton, J. D. (1993). General covariance and the foundations of general relativity: Eight decades of dispute, *Reports on Progress in Physics* **56**(7): 791–858.

Ogievetsky, V. I. and Polubarinov, I. V. (1965). Spinors in gravitation theory, *Soviet Physics, Journal of Experimental and Theoretical Physics* **21**: 1093–1104.

Oriti, D. (2009). The group field theory approach to quantum gravity, *in* D. Oriti (ed.), *Approaches to Quantum Gravity: Toward a New Understanding of Space, Time, and Matter*, Cambridge University Press, Cambridge.

Oriti, D. (2014). Disappearance and emergence of space and time in quantum gravity, *Studies in History and Philosophy of Modern Physics* **46**: 186–199.

Oriti, D. (2017). Group field theory and loop quantum gravity. **URL**: https://www.worldscientific.com/doi/10.1142/9789813220003_0005

Overduin, J. M. and Wesson, P. S. (1997). Kaluza-Klein gravity, *Physics Reports* **283**: 303–378.

Padmanabhan, T. (2008). From gravitons to gravity: Myth and reality, *International Journal of Modern Physics D* **17**(3): 367–398.

Page, B. (2018). Fine-tuned of necessity?, *Res Philosophica* **95**(4): 663–692.

Pasini, A. (1988). A conceptual introduction to Kaluza-Klein theory, *European Journal of Physics* **9**: 289–296.

Penrose, R. (1982). Some unsolved problems in classical general relativity, *in* S.-T. Yau (ed.), *Seminar on Differential Geometry*, Princeton University Press, Princeton.

Penrose, R. (2004). *The Road to Reality*, Jonathan Cape, London.

Percacci, R. (2009). Asymptotic safety, *in* D. Oriti (ed.), *Approaches to Quantum Gravity: Toward a New Understanding of Space, Time and Matter*, Cambridge University Press, Cambridge.

Pereira, J. G. (2014). Teleparallelism: A new insight into gravity, *in* A. Ashtekar and V. Petkov (eds), *Handbook of Spacetime*, Springer, Berlin.

Peskin, M. E. and Schroeder, D. V. (1995). *An Introduction to Quantum Field Theory*, Westview Press, London.

Pitts, J. B. (2006). Absolute objects and counterexamples: Jones-Geroch dust, Torretti constant curvature, tetrad-spinor, and scalar density, *Studies in History and Philosophy of Modern Physics* **37**(2): 347–371.

Pitts, J. B. (2009). Empirical equivalence, artificial gauge freedom and a generalized Kretschmann objection, *arXiv:0911.5400*.

Pitts, J. B. (2010). Gauge-invariant localization of infinitely many gravitational energies from all possible auxiliary structures, *General Relativity and Gravitation* **42**: 601–622.

Pitts, J. B. (2012). The nontriviality of trivial general covariance: How electrons restrict 'time' coordinates, spinors (almost) fit into tensor calculus, and $\frac{7}{16}$ of a tetrad is surplus structure, *Studies in History and Philosophy of Modern Physics* **43**: 1–24.

Pitts, J. B. (2016). Einstein's equations for spin 2 mass 0 from Noether's converse Hilbertian assertion, *Studies in History and Philosophy of Modern Physics* **56**: 60–69.

Pitts, J. B. and Schieve, W. C. (2007). Universally coupled massive gravity, *Theoretical and Mathematical Physics* **151**(2): 700–717.

Polchinski, J. (1998). *String Theory*, Vol. 1, Cambridge University Press, Cambridge.

Polchinski, J. (2017). Dualities of fields and strings, *Studies in History and Philosophy of Science Part B: Studies in History and Philosophy of Modern Physics* **59**: 6–20.

Pooley, O. (2002). *The Reality of Spacetime* D.Phil. thesis, University of Oxford.

Pooley, O. (2013). Substantivalist and relationist approaches to spacetime, *in* R. Batterman (ed.), *The Oxford Handbook of Philosophy of Physics*, Oxford University Press, Oxford.

Pooley, O. (2015). The reality of spacetime. Book manuscript.

Pooley, O. (2017). Background independence, diffeomorphism invariance, and the meaning of coordinates, *in* D. Lehmkuhl, G. Schiemann, and E. Scholz (eds), *Towards a Theory of Spacetime Theories*, Birkhäuser, Basel.

Pust, J. (2017). Intuition, *The Stanford Encyclopedia of Philosophy* Stanford University Press, Stanford. **URL**:https://plato.stanford.edu/archives/win2017/entries/intuition/

Read, J. (2016). The interpretation of string-theoretic dualities, *Foundations of Physics* **46**(2): 209–235.

Read, J. (2019a). Geometrical constructivism and modal relationalism: Further aspects of the dynamical/geometrical debate, *International Studies in the Philosophy of Science* **33**(1): 23–41.

Read, J. (2019b). On miracles and spacetime, *Studies in History and Philosophy of Modern Physics* **65**: 103–111.

Read, J. (2020). Functional gravitational energy, *British Journal for the Philosophy of Science* **71**(1): 205–232.

Read, J., Brown, H. R., and Lehmkuhl, D. (2018). Two miracles of general relativity, *Studies in History and Philosophy of Modern Physics* **64**: 14–25.

Read, J. and Le Bihan, B. (2021). The landscape and the multiverse: What's the problem? Synthese **199**: 7749–7771. **URL**: https://doi.org/10.1007/s11229-021-03137-0

Read, J. and Menon, T. (2021). The limitations of inertial frame spacetime functionalism, *Synthese* **199**(2): 229–251. **URL**: https://doi.org/10.1007/s11229-019-02299-2

Read, J. and Møller-Nielsen, T. (2020a). Motivating dualities, *Synthese* **197**: 263–291.

Read, J. and Møller-Nielsen, T. (2020b). Redundant epistemic symmetries, *Studies in History and Philosophy of Modern Physics* **70**: 88–97.

Read, J. and Teh, N. J. (2018). The teleparallel equivalent of Newton-Cartan gravity, *Classical and Quantum Gravity* **35**(18): 18LT01.

Read, J. and Teh, N. J. (2022). Newtonian equivalence principles, *Erkenntnis*. **URL**: https://doi.org/10.1007/s10670-021-00513-7

Reuter, M. and Weyer, H. (2009). The role of background independence for asymptotic safety in quantum Einstein gravity, *General Relativity and Gravitation* **41**: 983–1011.

Reutlinger, A. and Saatsi, J. (eds) (2018). *Explanation Beyond Causation: Philosophical Perspectives on Non-Causal Explanations*. Oxford University Press, Oxford.

Rickles, D. (2008a). Quantum gravity: A primer for philosophers, *The Ashgate Companion to Contemporary Philosophy of Physics*, Ashgate, London.

Rickles, D. (2008b). Who's afraid of background independence?, *The Ontology of Spacetime II. Vol. 4: Philosophy and Foundations of Physics*, Elsevier, London, pp. 133–152.

Rickles, D. (2011). A philosopher looks at string dualities, *Studies in History and Philosophy of Modern Physics* **42**: 54–67.

Rickles, D. (2013). Mirror symmetry and other miracles in superstring theory, *Foundations of Physics* **43**: 54–80.

Rosen, N. (1940a). General relativity and flat space, Vol. I, *Physical Review* **57**: 147–150.

Rosen, N. (1940b). General relativity and flat space, Vol. II, *Physical Review* **57**(2): 150–153.

Rosen, N. (1966). Flat space and variational principle, *in* B. Hoffmann (ed.), *Perspectives in Geometry and Relativity: Essays in Honor of Václav Hlavatý*, Indiana University Press, Indiana, ch. 33, pp. 325–327.

Rotger, B. C. (2014). Elementary features of Kaluza-Klein Theories. **URL**: https://diposit.ub.edu/dspace/bitstream/2445/59761/1/TFG_Cardona_Rotger_Biel.pdf

Rovelli, C. (1997). Halfway through the woods: Contemporary research on space and time, *in* J. Earman and J. D. Norton (eds), *The Cosmos of Science: Essays of Exploration. Vol. 6: Pittsburgh-Konstanz Series in the Philosophy and History of Science*, University of Pittsburgh Press, Pittsburgh, pp. 180–223.

Rovelli, C. (2001). Quantum spacetime: What do we know?, *in* C. Callender and N. Huggett (eds), *Physics Meets Philosophy at the Planck Scale: Contemporary Theories in Quantum Gravity*, Cambridge University Press, Cambridge, pp. 101–122.

Rovelli, C. (2004). *Quantum Gravity*, Cambridge University Press, Cambridge.

Rovelli, C. (2010). A new look at loop quantum gravity, *Classical and Quantum Gravity* **28**: 1–24.

Rovelli, C. (2013). A critical look at strings, *Foundations of Physics* **43**(1): 8–20.

Rovelli, C. and Vidotto, F. (2015). *Covariant Loop Quantum Gravity: An Elementary Introduction to Quantum Gravity and Spinfoam Theory*, Cambridge University Press.

Rozali, M. (2009). Comments on background independence and gauge redundancies, *Advanced Science Letters* **2**(2): 244–250.

Samaroo, R. (2011). On identifying background-structure in classical field theories, *Philosophy of Science* **78**: 1070–1081.

Saunders, S. (2003). Physics and Leibniz's principles, *in* K. Brading and E. Castellani (eds), *Symmetries in Physics: Philosophical Reflections*, Cambridge University Press, Cambridge, pp. 151–168.

Saunders, S. (2007). Mirroring as an a priori symmetry, *Philosophy of Science* **74**(4): 452–480.

Saunders, S. (2013). Rethinking Newton's *Principia, Philosophy of Science* **80**: 22–48.

Schouten, J. A. (1954). *Ricci-Calculus: An Introduction to Tensor Analysis and Its Geometrical Applications*, Springer, Berlin.

Schuller, F. (2016). Lectures on measure theory, **URL**: https://www.youtube.com/watch?v=6ad9V8gvyBQ.

Schwartz, P. K. (2022). Teleparallel Newton–Cartan Gravity, *Classical and Quantum Gravity* **40**: 105008.

Seiberg, N. (2007). Emergent spacetime, *in* D. Gross, M. Henneaux, and A. Sevrin (eds), *The Quantum Structure of Space and Time*, World Scientific, London.

Sloan, D. (2022). Herglotz action for homogeneous cosmology. Unpublished manuscript.

Smolin, L. (2003). Time, structure and evolution in cosmology, *in* A. Ashtekar, R. S. Cohen, D. Howard, J. Renn, S. Sarkar, and A. Shimony (eds), *Revisiting the Foundations of Relativistic Physics: Festschrift in Honor of John Stachel*. Vol. 234: *Boston Studies in the Philosophy of Science*, Kluwer, Dordrecht, pp. 221–274.

Smolin, L. (2006). The case for background independence, *in* D. Rickles, S. French, and J. Saatsi (eds), *The Structural Foundations of Quantum Gravity*, Oxford University Press, Oxford, pp. 196–239.

Smolin, L. (2008). *The Trouble with Physics: The Rise of String Theory, The Fall of a Science, and What Comes Next*, Penguin, London.

Smolin, L. (2013). A perspective on the landscape problem, *Foundations of Physics* **43**: 21–45.

Sorkin, R. (2002). An example relevant to the Kretschmann-Einstein debate, *Modern Physics Letters A* **17**: 695–700.

Stewart, J. (1991). *Advanced General Relativity*, Cambridge University Press, Cambridge.

Suppes, P. (1960). A comparison of the meaning and uses of models in mathematics and the empirical sciences, *Synthese* **12**: 287–301. **URL**: https://doi.org/10.1007/BF00485107.

Sus, A. (2008). *General Relativity and the Physical Content of General Covariance*, PhD thesis, Universitat Aut'onoma de Barcelona.

Sus, A. (2010). Absolute objects and general relativity: Dynamical considerations, *in* M. Suarez, M. Dorato, and M. R'edei (eds), *EPSA Philosophical Issues in the Sciences: Launch of the European Philosophy of Science Association*, Vol. 2, Springer, Berlin, pp. 239–249.

't Hooft, G. (1993). Dimensional reduction in quantum gravity. **URL**: https://arxiv.org/abs/gr-qc/9310026

Teh, N. J. (2013). Holography and emergence, *Studies in History and Philosophy of Modern Physics* **44**: 300–311.

Teh, N. J. (2016). Galileo's gauge: Understanding the empirical significance of gauge symmetry, *Philosophy of Science* **83**(1): 93–118.

Teh, N. J. (2018). Recovering recovery: On the relationship between gauge symmetry and Trautman recovery, *Philosophy of Science* **85**(1): 201–224.

Teitel, T. (2019). Background independence: Lessons for further decades of dispute, *Studies in History and Philosophy of Modern Physics* **65**: 41–54.

Thorne, K. S., Lee, D. L., and Lightman, A. P. (1973). Foundations for a theory of gravitation theories, *Physical Review D* **7**: 3563–3578.

Torretti, R. (1984). Space-time physics and the philosophy of science: Review of *Foundations of Space-Time Theories: Relativistic Physics and the Philosophy of Science* by Michael Friedman, *British Journal for the Philosophy of Science* **35**: 280–292.

Trautman, A. (1962). Conservation laws in general relativity, *in* L. Witten (ed.), *Gravitation: Introduction to Current Research*, Wiley, New York, pp. 169–198.

Trautman, A. (1965). Foundations and current problems of general relativity, in A. Trautmann, F. A. E. Pirani, and H. Bondi (eds.), *Lectures on General Relativity*, Englewood Cliffs, NJ: Prentice-Hall Inc., pp. 1–248.

Trautman, A. (2006). Einstein-Cartan theory, *in* J.-P. Francoise, G. L. Naber, and T. S. T. (eds), *Encyclopedia of Mathematical Physics*, Elsevier, London pp. 189–195.

van Fraassen, B. C. (1980). *The Scientific Image*, Oxford University Press, Oxford.

van Fraassen, B. C. (1989). *Laws and Symmetry*, Oxford University Press, Oxford.

van Fraassen, B. C. (2002). *The Empirical Stance*, Yale University Press, New Haven, CN.

Vassallo, A. (2015). General covariance, diffeomorphism invariance, and background independence in 5 dimensions, *in* T. Bigaj and C. Wüthrich (eds), *Metaphysics in Contemporary Physics*, Rodopi, Amsterdam.

Vistarini, T. (2019). *The Emergence of Spacetime in String Theory*, Routledge, London.

Wald, R. (1984). *General Relativity*, University of Chicago Press, Chicago.

Wallace, D. (2006). In defence of naiveté: The conceptual status of Lagrangian quantum field theory, *Synthese* **151**: 33–80.

Wallace, D. (2012). *The Emergent Multiverse: Quantum Theory According to the Everett Interpretation*, Oxford University Press, Oxford.

Wallace, D. (2015). Fields as bodies: A unified presentation of spacetime and internal gauge symmetry. **URL**: http://philsci-archive.pitt.edu/11337/

Wallace, D. (2019). Who's afraid of coordinate systems? An essay on representation of spacetime structure, *Studies in History and Philosophy of Modern Physics* **67**: 125–136.

Wallace, D. (2020). Fundamental and emergent geometry in Newtonian physics, *British Journal for the Philosophy of Science* **71**: 1–32.

Wallace, D. (2022a). Isolated systems and their symmetries. Part I: General framework and particle-mechanics examples, *Studies in History and Philosophy of Science* **92**: 239–248.

Wallace, D. (2022b). Observability, redundancy and modality for dynamical symmetry transformations, *in* J. Read and N. Teh (eds), *The Philosophy and Physics of Noether's Theorems*, Cambridge University Press, Cambridge.

Wallace, D. and Timpson, C. G. (2010). Quantum mechanics on spacetime. Part I: Spacetime state realism, *British Journal for the Philosophy of Science* **61**(4): 697–727.

Weatherall, J. O. (2016a). Are Newtonian gravitation and geometrized Newtonian gravitation theoretically equivalent?, *Erkenntnis* **81**: 1073–1091.

Weatherall, J. O. (2016b). Fiber bundles, Yang-Mills theory, and general relativity, *Synthese* **193**: 2389–2425.

Weatherall, J. O. (2016c). Maxwell-Huygens, Newton-Cartan, and Saunders-Knox spacetimes, *Philosophy of Science* **83**: 82–92.

Weatherall, J. O. (2016d). Understanding gauge, *Philosophy of Science* **85**: 1039–1049.

Weatherall, J. O. (2017). Categories and the foundations of classical field theories, *in* E. Landry (ed.), *Categories for the Working Philosopher*, Oxford University Press, Oxford, pp. 329–348.

Weatherall, J. O. (2018). A brief comment on Maxwell(/Newton)[-Huygens] spacetime, *Studies in History and Philosophy of Modern Physics* **63**: 34–38.

Weatherall, J. O. (2019a). Theoretical equivalence in physics, Part I, *Philosophy Compass* **14**.

Weatherall, J. O. (2019b). Theoretical equivalence in physics, Part II, *Philosophy Compass* **14**.

Weatherall, J. O. (2020). Equivalence and duality in electromagnetism, *Philosophy of Science* **87**: 1172–1183.

Weatherall, J. O. (2021). Why not categorical equivalence?, *in* J. Madarász and G. Székely (eds), *Hajnal Andréka and István Németi on Unity of Science: From Computing to Relativity Theory Through Algebraic Logic*, Springer, Berlin, pp. 427–451.

Williamson, T. (2007). *The Philosophy of Philosophy*, John Wiley & Sons, London.

Witten, E. (1995). Some comments on string dynamics, *arXiv preprint hep-th/9507121.*

Witten, E. (1996). Reflections on the fate of spacetime, *Physics Today* **49**(4): 24–30.

Wolf, W. J. and Read, J. (2023). Respecting boundaries: Theoretical equivalence and structure beyond dynamics, *arXiv preprint arXiv:2302.07180.*

Wolf, W. J., Read, J., and Teh, N. J. (2022). Edge modes and dressing fields for the Newton-Cartan quantum Hall effect, *Foundations of Physics* **53**(1): 1–24.

Wolf, W. J., Sanchioni, M., and Read, J. (2023). Underdetermination in classic and modern tests of general relativity. Unpublished manuscript.

Wüthrich, C. (2006). *Approaching the Planck Scale from a Generally Relativistic Point of View: A Philosophical Appraisal of Loop Quantum Gravity*, PhD thesis, University of Pittsburgh.

Wüthrich, C. (2017). Raiders of the lost spacetime, *in* D. Lehmkuhl, G. Schiemann, and E. Scholz (eds), *Towards a Theory of Spacetime Theories*, Birkhäuser, Basel.

Index